OCEANS

OCEANS

A SCIENTIFIC AMERICAN READER

THE UNIVERSITY OF CHICAGO PRESS · CHICAGO AND LONDON

The University of Chicago Press, Chicago 60637
The University of Chicago Press, Ltd., London
© 2007 by Scientific American, Inc.
All rights reserved. Published 2007
Printed in the United States of America

16 15 14 13 12 11 10 09 08 07   1 2 3 4 5

ISBN-13: 978-0-226-74260-1 (cloth)
ISBN-13: 978-0-226-74262-5 (paper)
ISBN-10: 0-226-74260-1 (cloth)
ISBN-10: 0-226-74262-8 (paper)

Library of Congress Cataloging-in-Publication Data

Oceans : a Scientific American reader / Scientific American.
    p.   cm.
  ISBN-13: 978-0-226-74260-1 (cloth : alk. paper)
  ISBN-13: 978-0-226-74262-5 (pbk. : alk. paper)
  ISBN-10: 0-226-74260-1 (cloth : alk. paper)
  ISBN-10: 0-226-74262-8 (pbk. : alk. paper)
    1. Ocean.   I. Scientific American, inc.

GC21. 028 2007
551.46—dc22

                                    2006015664

CONTENTS

ORIGINS

Sculpting the Earth from Inside Out

MICHAEL GURNIS

ORIGINALLY PUBLISHED MARCH 2001

Credit for sculpting the earth's surface typically goes to violent collisions between tectonic plates, the mobile fragments of the planet's rocky outer shell. The mighty Himalayas shot up when India rammed into Asia, for instance, and the Andes grew as the Pacific Ocean floor plunged beneath South America. But even the awesome power of plate tectonics cannot fully explain some of the planet's most massive surface features.

Take southern Africa. This region boasts one of the world's most expansive plateaus, more than 1,000 miles across and almost a mile high. Geologic evidence shows that southern Africa, and the surrounding ocean floor, has been rising slowly for the past 100 million years, even though it has not experienced a tectonic collision for nearly 400 million years.

The African superswell, as this uplifted landmass is known, is just one example of dramatic vertical movement by a broad chunk of the earth's surface. In other cases from the distant past, vast stretches of Australia and North America bowed down thousands of feet—and then popped up again.

Scientists who specialize in studying the earth's interior have long suspected that activity deep inside the earth was behind such vertical changes at the surface. These geophysicists began searching for clues in the mantle—the middle layer of the planet. This region of scalding-hot rock lies just below the jigsaw configuration of tectonic plates and extends down more than 1,800 miles to the outer edge of the globe's iron core. Researchers learned that variations in the mantle's intense heat and pressure enable the solid rock to creep molasseslike over thousands of years. But they could not initially decipher how it could give rise to large vertical motions. Now, however, powerful computer models that combine snapshots of the mantle today with clues about how it might have behaved in the past are beginning to explain why parts of the earth's surface have undergone these astonishing ups and downs.

The mystery of the African superswell was among the easiest to decipher. Since the early half of the 20th century, geophysicists have understood that over the unceasing expanse of geologic time, the mantle not only creeps, it churns and roils like a pot of thick soup about to boil. The relatively low density of the hottest rock makes that material buoyant, so it ascends slowly; in contrast, colder, denser rock sinks until heat escaping the molten core warms it enough to make it rise again. These three-dimensional motions, called convection, are known to enable the horizontal movement of tectonic plates, but it seemed unlikely that the forces they created could lift and lower the planet's surface. That skepticism about the might of the mantle began to fade away when researchers created the first blurry images of the earth's interior.

About 20 years ago scientists came up with a way to make three-dimensional snapshots of the mantle by measuring vibrations that are set in motion by earthquakes originating in the planet's outer shell. The velocities of these vibrations, or seismic waves, are determined by the chemical composition, temperature and pressure of the rocks they travel through. Waves become sluggish in hot, low-density rock, and they speed up in colder, denser regions. By recording the time it takes for seismic waves to travel from an earthquake's epicenter to a particular recording station at the surface, scientists can infer the temperatures and densities in a given segment of the interior. And by compiling a map of seismic velocities from thousands of earthquakes around the globe they can begin to map temperatures and densities throughout the mantle.

These seismic snapshots, which become increasingly more detailed as researchers find more accurate ways to compile their measurements, have recently revealed some unexpectedly immense formations in the deepest parts of the mantle. The largest single structure turns out to lie directly below Africa's southern tip. About two years ago seismologists Jeroen Ritsema and Hendrik-Jan van Heijst of the California Institute of Technology calculated that this mushroom-shaped mass stretches some 900 miles upward from the core and spreads across several thousand miles.

The researchers immediately began to wonder whether this enormous blob could be shoving Africa skyward. Because the blob is a region where seismic waves are sluggish, they assumed that it was hotter than the surrounding mantle. The basic physics of convection suggested that a hot blob was likely to be rising. But a seismic snapshot records only a single moment in time and thus only one position of a structure. If the blob were of a different composition than the surrounding rock, for instance, it could be hotter and still not rise. So another geophysicist, Jerry X. Mitrovica of the University of Toronto, and I decided to create a time-lapse picture of what might be happening. We plugged the blob's shape and estimated density, along with estimates of when southern Africa began rising, into a computer program that simulates mantle convection. By doing so, we found last year that the blob is indeed buoyant enough to rise slowly within the mantle—and strong enough to push Africa upward as it goes.

Seismic snapshots and computer models—the basic tools of geophysicists—were enough to solve the puzzle of the African superswell, but resolving the up-and-down movements of North America and Australia was more complicated and so was accomplished in a more circuitous way. Geophysicists who think only about what the mantle looks like today cannot fully explain how it sculpts the earth's surface. They must therefore borrow from the historical perspective of traditional geologists who think about the way the surface has changed over time.

GHOSTS FROM THE PAST

The insights that would help account for the bobbings of Australia and North America began to emerge with investigations of a seemingly unrelated topic: the influence of mantle density on the earth's gravitational field. The basic principles of physics led scientists in the 1960s to expect that gravity would be lowest above pockets of hot rock, which are less dense and thus have less mass. But when geophysicists first mapped the earth's gravitational variations, they found no evidence that gravity correlated with the cold and hot parts of the mantle—at least not in the expected fashion.

Indeed, in the late 1970s and early 1980s Clement G. Chase uncovered the opposite pattern. When Chase, now at the University of Arizona, considered geographic scales of more than 1,000 miles, he found that the pull of gravity is strongest not over cold mantle but over isolated volcanic regions called hot spots. Perhaps even more surprising was what Chase noticed about the position of a long band of low gravity that passes from Hudson Bay in Canada northward over the North Pole, across Siberia and India, and down into Antarctica. Relying on estimates of the ancient configuration of tectonic plates, he showed that this band of low gravity marked the location of a series of subduction zones—that is, the zones where tectonic plates carrying fragments of the seafloor plunge back into the mantle—from 125 million years ago. The ghosts of ancient subduction zones seemed to be diminishing the pull of gravity. But if cold, dense chunks of seafloor were still sinking through the mantle, it seemed that gravity would be high above these spots, not low, as Chase observed.

In the mid-1980s geophysicist Bradford H. Hager, now at the Massachusetts Institute of Technology, resolved this apparent paradox by proposing that factors other than temperature might create pockets of extra or deficient mass within the mantle. Hager developed his theory from the physics that describe moving fluids, whose behavior the mantle imitates over the long term. When a low-density fluid rises upward, as do the hottest parts of the mantle, the force of the flow pushes up the higher-density fluid above it. This gentle rise atop the upwelling itself creates an excess of mass (and hence stronger gravity) near the planet's surface. By the same token, gravity can be lower over cold, dense material: as this heavy matter sinks, it drags down mass that was once near the surface. This conception explained why the ghosts of subduction zones could generate a band of low gravity: some of that cold, subducted seafloor must still be sinking within the mantle—and towing the planet's surface downward in the process. If Hager's explanation was correct, it meant that the mantle did not merely creep horizontally near the planet's surface; whole segments of its up-and-down movements also reached the surface. Areas that surged upward would push the land above it skyward, and areas that sank would drag down the overlying continents as they descended.

BOBBING CONTINENTS

At the same time that Chase and Hager were discovering a mechanism that could dramatically lift and lower the earth's surface, geologists were beginning to see evidence that continents might actually have experienced such dips and swells in the past. Geologic formations worldwide contain evidence that sea level fluctuates over time. Many geologists suspected that this fluctuation would affect all continents in the same way, but a few of them advanced convincing evidence that the most momentous changes in sea level stemmed from vertical motions of continents. As one continent moved, say, upward relative to other landmasses, the ocean surface around that continent would become lower while sea level around other landmasses would stay the same.

Most geologists, though, doubted the controversial notion that continents could move vertically—even when the first indications of the bizarre bobbing of Australia turned up in the early 1970s. Geologist John J. Veevers of Macquarie University in Sydney examined outcrops of ancient rock in eastern Australia and discovered that sometime in the early Cretaceous period (about 130 million years ago), a shallow sea rapidly covered that half of Australia while other continents flooded at a much more leisurely pace. Sea level climaxed around those landmasses by the late Cretaceous (about 70 million years ago), but by then the oceans were already retreating from Australia's shores. The eastern half of the continent must have sunk several thousand feet relative to other landmasses and then popped back up before global sea level began to fall.

Veevers's view of a bobbing continent turned out to be only part of Australia's enigmatic story. In 1978 geologist Gerard C. Bond, now at Columbia University's Lamont-Doherty Earth Observatory, discovered an even stranger turn of events while he was searching global history for examples of vertical continental motion. After Australia's dip and rise during the Cretaceous, it sank again, this time by 600 feet, between then and the present day. No reasonable interpretation based on plate tectonics alone could explain the widespread vertical motions that Bond and Veevers uncovered. Finding a satisfactory explanation would require scientists to link this information with another important clue: Hager's theory about how the mantle can change the shape of the planet's surface.

The first significant step in bringing these clues together was the close examination of another up-and-down example from Bond's global survey. In the late 1980s this work inspired Christopher Beaumont, a geologist at Dalhousie University in Nova Scotia, to tackle a baffling observation about Denver, Colo. Although the city's elevation is more than a mile above sea level, it sits atop flat, undeformed marine rocks created from sediments deposited on the floor of a shallow sea during the Cretaceous period. Vast seas covered much of the continents during that time, but sea level was no more than about 400 feet higher than it is today. This means that the ocean could never have reached as far inland as Denver's current position—unless this land was first pulled down several thousand feet to allow waters to flood inland.

Based on the position of North America's coastlines during the Cretaceous, Beaumont estimated that this bowing downward and subsequent uplift to today's elevation must have affected an area more than 600 miles across. This geographic scale was problematic for the prevailing view that plate tectonics alone molded the surface. The mechanism of plate tectonics permits vertical motions within only 100 miles or so of plate edges, which are thin enough to bend like a stiff fishing pole, when forces act on them. But the motion of North America's interior happened several hundred miles inland—far from the influence of plate collisions. An entirely different mechanism had to be at fault.

Beaumont knew that subducted slabs of ancient seafloor might sit in the mantle below North America and that such slabs could theoretically drag down the center of a continent. To determine whether downward flow of the mantle could have caused the dip near Denver, Beaumont teamed up with Jerry Mitrovica, then a graduate student at the University of Toronto, and Gary T. Jarvis of York University in Toronto. They found that the sinking of North America during the Cretaceous could have been caused by a plate called the Farallon as it plunged into the mantle beneath the western coast of North America. Basing their conclusion on a computer model, the research team argued that the ancient plate thrust into the mantle nearly horizontally. As it began sinking, it created a downward flow in its wake that tugged North America low enough to allow the ocean to rush in. As the Farallon plate sank deeper, the power of its trailing wake decreased. The continent's tendency to float eventually won out, and North America resurfaced.

When the Canadian researchers advanced their theory in 1989, the Farallon plate had long since vanished into the mantle, so its existence had only been inferred from geologic indications on the bottom of the Pacific Ocean. At that time, no seismic images were of high enough resolution to delineate a structure as small as a sinking fragment of the seafloor. Then, in 1996, new images of the mantle changed everything. Stephen P. Grand of the University of Texas at Austin and Robert D. van der Hilst of M.I.T., seismologists from separate research groups, presented two images based on entirely different sets of seismic measurements. Both pictures showed virtually identical structures, especially the cold-mantle downwellings associated with sinking slabs of seafloor. The long-lost Farallon plate was prominent in the images as an arching slab 1,000 miles below the eastern coast of the U.S.

MOVING DOWN UNDER

Connecting the bobbing motion of North America to the subduction of the seafloor forged a convincing link between ancient sea-level change and goings-on in the mantle. It also became clear that the ancient Farallon slab sits within the band of low gravity that Chase had observed two decades earlier. I suspected that these ideas could also be applied to the most enigmatic of the continental bobbings, that of Australia during and since the Cretaceous. I had been simulating mantle convection with computer models for 15 years, and many of my results showed that the mantle was in fact able to lift the surface by thousands of feet—a difference easily great enough to cause an apparent drop in sea level. Like Chase, Veevers and other researchers before me, I looked at the known history of plate tectonics for clues about whether something in the mantle could have accounted for Australia's bouncing. During the Cretaceous period, Australia, South America, Africa, India, Antarctica and New Zealand were assembled into a vast supercontinent called Gondwana, which had existed for more than 400 million years before it fragmented into today's familiar landmasses. Surrounding Gondwana for most of this time was a huge subduction zone where cold oceanic plates plunged into the mantle.

I thought that somehow the subduction zone that surrounded Gondwana for hundreds of millions of years might have caused Australia's ups and downs. I became more convinced when I sketched the old subduction zones on maps of ancient plate configurations constructed by R. Dietmar Müller, a seagoing geophysicist at Sydney University. The sketches seemed to explain the Australian oddities. Australia would have passed directly over Gondwana's old subduction zone at the time it sank.

To understand how the cold slab would behave in the mantle as Gondwana broke apart over millions of years, Müller and I joined Louis Moresi of the Commonwealth Scientific and Industrial Research Organization in Perth to run a computer simulation depicting the mantle's influence on Australia over time. We knew the original position of the ancient subduction zone, the history of horizontal plate motions in the region and the estimated properties—such as viscosity—of the mantle below. Operating under these constraints, the computer played out a scenario for Australia that fit our hypotheses nearly perfectly.

The computer model started 130 million years ago with ocean floor thrusting beneath eastern Australia. As Australia broke away from Gondwana, it passed over the cold, sinking slab, which sucked the Australian plate downward. The continent rose up again as it continued its eastward migration away from the slab.

Our model resolved the enigma of Australia's motion during the Cretaceous, originally observed by Veevers, but we were still puzzled by the later continentwide sinking of Australia that Bond discovered. With the help of another geophysicist, Carolina Lithgow-Bertelloni, now at the University of Michigan, we confirmed Bond's observation that as Australia moved northward toward Indonesia after the Cretaceous, it subsided by about 600 feet. Lithgow-Bertelloni's global model of the mantle, which incorporated the history of subduction, suggested that Indonesia is sucked down more than any other region in the world because it lies at the intersection of enormous, present-day subduction systems in the Pacific and Indian oceans. And as Indonesia sinks, it pulls Australia down with it.

Today Indonesia is a vast submerged continent—only its highest mountain peaks protrude above sea level.

Which brings us back to Africa. In a sense, Indonesia and Africa are opposites: Indonesia is being pulled down while Africa is being pushed up. These and other changes in the mantle that have unfolded over the past few hundred million years are intimately related to Gondwana. The huge band of low gravity that Chase discovered 30 years ago is created by the still-sinking plates of a giant subduction zone that once encircled the vast southern landmass. At the center of Gondwana was southern Africa, which means that the mantle below this region was isolated from the chilling effects of sinking tectonic plates at that time—and for the millions of years since. This long-term lack of cold, downward motion below southern Africa explains why a hot superplume is now erupting in the deep mantle there.

With all these discoveries, a vivid, dynamic picture of the motions of the mantle has come into focus. Researchers are beginning to see that these motions sculpt the surface in more ways than one. They help to drive the horizontal movement of tectonic plates, but they also lift and lower the continents. Perhaps the most intriguing discovery is that motion in the deep mantle lags behind the horizontal movement of tectonic plates. Positions of ancient plate boundaries can still have an effect on the way the surface is shaped many millions of years later.

Our ability to view the dynamics of mantle convection and plate tectonics will rapidly expand as new ways of observing the mantle and techniques for simulating its motion are introduced. When mantle convection changes, the gravitational field changes. Tracking variations in the earth's gravitational field is part of a joint U.S. and German space mission called GRACE, which is set for launch in June. Two spacecraft, one chasing the other in earth orbit, will map variations in gravity every two weeks and perhaps make it possible to infer the slow, vertical flow associated with convection in the mantle. Higher-resolution seismic images will also play a pivotal role in revealing what the mantle looks like today. Over the five- to 10-year duration of a project called USArray, 400 roving seismometers will provide a 50-mile-resolution view into the upper 800 miles of the mantle below the U.S.

Plans to make unprecedented images and measurements of the mantle in the coming decade, together with the use of ever more powerful supercomputers, foretell an exceptionally bright future for deciphering the dynamics of the earth's interior. Already, by considering the largest region of the planet—the mantle—as a chunk of rock with a geologic history, earth scientists have made extraordinary leaps in understanding the ulti-mate causes of geologic changes at the surface.

FURTHER READING

DYNAMICS OF CRETACEOUS VERTICAL MOTION OF AUSTRALIA AND THE AUSTRALIAN-ANTARCTIC DISCORDANCE. Michael Gurnis, R. Dietmar Müller and Louis Moresi in *Science,* Vol. 279, pages 1499–1504; March 6, 1998.

DYNAMIC EARTH: PLATES, PLUMES AND MANTLE CONVECTION. Geoffrey F. Davies. Cambridge University Press, 2000.

CONSTRAINING MANTLE DENSITY STRUCTURE USING GEOLOGICAL EVIDENCE OF SURFACE UPLIFT RATES: THE CASE OF THE AFRICAN SUPERPLUME. Michael Gurnis, Jerry X. Mitrovica, Jeroen Ritsema and Hendrik-Jan van Heijst in *Geochemistry, Geophysics, Geosystems,* Vol. 1, Paper No. 1999GC000035; 2000. Available online at http://146.201.254.53/publicationsfinal/articles/1999GC000035/fs1999GC000035.html.

Gurnis's Computational Geodynamics Research Group Web site: www.gps.caltech.edu/~gurnis/geodynamics.html

Large Igneous Provinces

MILLARD F. COFFIN AND OLAV ELDHOLM

ORIGINALLY PUBLISHED OCTOBER 1993

Sometimes getting there is not half the fun of geophysical research. Bearing westward into a rising Southern Ocean gale, the Australian research vessel *Rig Seismic* was pounded so hard that its anchor broke loose and smashed into the forecastle, causing the waves breaking over the bow to flood parts of the ship. Progress toward the Kerguelen Plateau, an elevated region of seafloor just north of Antarctica, was delayed while the ship's crew secured the anchor and welded the ship to keep it watertight. Several days later the expedition, including one of us (Coffin), arrived at Kerguelen, deployed instruments for analyzing the structure of the ocean bottom and began collecting data. The aim of the voyage was to understand the origin and evolution of the huge underwater plateau.

In that same year, 1985, but in the other hemisphere, a floating earth science laboratory and drilling vessel, *JOIDES Resolution*, steamed north through stormy North Atlantic seas to the Vøring Plateau off the coast of Norway. That expedition, led by one of us (Eldholm) and Jörn Thiede of the Research Center for Maritime Earth Sciences in Kiel, Germany, had as its purpose the investigation of the geologic structures that form when continents tear apart and an ocean is born. Confounding the expectations of most of their colleagues, the researchers managed to obtain core samples of igneous rock nearly one kilometer deep, lying under 300 meters of soft sediment.

Although they lie thousands of kilometers apart in disparate geologic settings, we recognized fundamental similarities in the two plateaus. Our findings, merged with seismic and drilling data gathered by many other workers, have helped demonstrate that the Kerguelen Plateau and the volcanic continental margin off Norway belong to a class of huge magmatic features generically known as large igneous provinces. These features cover areas of up to millions of square kilometers, and yet they seem to have formed quite swiftly in geologic terms.

The spasms of eruptive activity associated with such rapid outpourings of lava may have substantially affected the chemistry and circulation of the oceans and atmosphere. Some of the resulting environmental shifts may have contributed to mass extinctions, including the one in which the dinosaurs vanished; in contrast, other changes may have promoted biological diversity and the origin of new species.

The recognition of large igneous provinces has forced geophysicists to rethink their notions regarding the structure of the earth's interior. The theory of plate tectonics tidily accounts for the slow, steady volcanic activity that occurs at mid-ocean ridges (where new oceanic crust is born) and near subduction zones (where old, dense sections of the ocean floor sink back into the earth's hot interior), but it cannot readily explain the abrupt outbursts necessary to create large igneous provinces. Although the rocks that make up these features generally resemble the composition of the lavas that emerge at mid-ocean ridges, igneous provinces differ in their trace element content and mix of atomic isotopes. And they sometimes break through the middle of the normally placid lithospheric plates.

In the 1960s the late J. Tuzo Wilson of the University of Toronto, followed by W. Jason Morgan of Princeton University, developed a hypothesis that helped to elucidate the phenomenon. The researchers proposed that the earth's mantle, the vast zone lying below the crust but above the core, circulates in two different modes. The dominant one consists of large-scale convection that nudges plates across the surface and causes continents to drift. But about one tenth of the heat now escaping from the mantle does so in the form of deep-rooted, narrow plumes of warmer than average material that rise through the mantle. When it reaches the base of the lithosphere, a plume decompresses and partially melts, producing an upwelling of magma and a long-lasting locus of volcanic activity known as a hot spot. Some of that magma may erupt as a tremendous flood of lava. Unlike the slow, steady drifting of continents and the spreading of mid-ocean ridges, the surfacing of mantle plumes takes place in an erratic and episodic manner.

All models of the internal structure of the earth are built on inference. Even the deepest boreholes extend only about 10 kilometers. Direct and indirect studies of large igneous provinces are therefore vitally important for learning more about the nature of mantle plumes and how they may ultimately affect conditions on the earth's surface.

The most fundamental observation about these provinces is that they consist of basalt, a common, iron- and magnesium-rich rock. The large igneous features that appear in the middle of continents, where geologists can easily collect samples and determine their composition, are therefore referred to as continental flood basalts. At the edges of continents, igneous provinces are called volcanic passive margins; if they are located in the middle of the ocean, they are dubbed oceanic plateaus.

Geologists recognized the existence of continental flood basalts late in the 19th century, when they realized that several far-flung volcanic constructions actually constitute connected flows of basaltic lava. One of the most spectacular of these is the Deccan Traps in west-central India. (*Trap* is Dutch for "staircase," a reference to the steplike appearance of the eroded lavas.) Similar structures include the Columbia River basalts in the northwestern U.S.; the North Atlantic volcanic outpourings of the British Isles, the Faeroe Islands and Greenland; and the Karoo basalts in southern Africa, to name just a few. The layers of lava in a flood basalt pile up several kilometers thick. Individual flows may contain several thousand cubic kilometers of rock and may extend hundreds of kilometers.

Because of the inaccessibility of underwater igneous provinces, earth scientists only recently became aware of their resemblance to continental flood basalts. In 1981 Karl Hinz of the Federal Institute for Geosciences and Natural Resources in Hannover, Germany, suggested, based on his analysis of recordings of reflected seismic waves, that the submerged margins of many continents contain extensive, layered lava flows. Since then, investigators have obtained improved seismic data that seem to corroborate Hinz's conclusion. Igneous rocks recovered by scientific drilling projects on the continental margins off Ireland and Norway confirm that the structures seen in the seismic images do in fact have volcanic origins.

Analogous seismic studies of the Kerguelen Plateau present convincing evidence that oceanic plateaus likewise consist primarily of volcanic rocks. More recently scientists on board the *JOIDES Resolution* have recovered direct samples of rocks from two immense oceanic plateaus: Kerguelen in 1988 and Ontong Java, situated in the Pacific Ocean northeast of Australia, in 1990. The samples consist of basaltic rock, similar to that found in continental flood basalts.

Earth scientists have wondered how such massive volcanic constructions arise. Unfortunately, the most basic information needed to answer that question—the total volumes of lava and intrusive rock in igneous provinces and the rate at which they formed—is poorly known. On land, geologists can directly measure the area of flood basalts but can only infer their depth. The crust under continents averages 35 kilometers in thickness. Only the top parts are accessible to exploration. Beneath the oceans the situation is even more difficult. Sampling just the uppermost sections of the five- to 25-kilometer-thick crust in which oceanic plateaus reside requires the services of an expensive drilling vessel.

The passage of time also disguises the true extent of ancient volcanic structures. The older igneous provinces are often heavily eroded and are therefore both disfigured and diminished in scale. John J. Mahoney of the University of Hawaii estimates that the Deccan Traps, for example, originally covered an area three times larger than at present. Moreover, in the 65 million years since the formation of the Deccan Traps, seafloor spreading has apparently dispersed sections of the original lavas to the Seychelles, the Mascarene Plateau and the Chagos-Laccadive Ridge far to the south and southwest of India.

Undersea volcanic margins and oceanic plateaus similarly have been reconfigured by the forces of change. Drill samples and seismic-reflection images of the igneous province along the Norwegian continental margin and of the Kerguelen Plateau demonstrate that when those formations first erupted, they stood above sea level and only gradually subsided to their present position in deep water. Coffin has estimated that the Kerguelen and Broken Ridge plateaus, along with most of the other oceanic plateaus and ridges in the Indian Ocean, formed above sea level and remained there for as long as 50 million years, an ample span of time in which erosion could proceed.

Despite the paucity of data and the physical transformations of large igneous provinces, we have managed to deduce the original areas covered by lava for five of the best studied of these formations: the Ontong Java and Kerguelen-Broken Ridge oceanic plateaus, the North Atlantic volcanic margins and the Deccan and Columbia River continental flood basalts. The smallest of these—the Columbia River basalts—encompasses a region larger than the state of New York; the largest, the Ontong Java, is two thirds the size of Australia. To evaluate the area and volume of large igneous provinces, geologists realized that they needed to examine modern-day analogues of the gigantic igneous provinces.

No currently active volcanic regions even begin to approach the extinct large igneous provinces in magnitude. The most active modern hot spots (the big island of Hawaii and Réunion, an island lying to the east of Madagascar) cover an area of the earth's crust only about one fifth the size of the Columbia River basalts. Nevertheless, studies of the crustal structure below Hawaii contribute to the understanding of the generic forms that underlie all volcanic hot spots.

In 1982 Anthony B. Watts and Uri S. ten Brink and their co-workers at the Lamont-Doherty Earth Observatory collected seismic data by recording artificially generated sound waves that are reflected and refracted underneath the Hawaiian Islands and seamounts. The speed of those waves depends on the physical properties (such as density and elastic constants) of the rock; where those parameters change, the seismic waves will alter course. Seismic analysis reveals that the Hawaiian Islands built up as basaltic magma intruded into and collected atop the preexisting oceanic crust.

Those surface lavas evidently do not contain all of the material associated with the islands' formation, however. Below the islands and seamounts lies an anomalous zone of rock distinguished by the rapid rate at which compressional waves travel through it. Geologists think the underlying body of rock derived from the same mantle source as did the islands and seamounts. Existing techniques cannot reveal how much new igneous material was added to the previously existing oceanic crust as the islands formed. Any calculated volumes of mantle plume–derived material contained in the Hawaiian Chain therefore must be considered minimal.

Even so, we were able to establish a basic relation between seismic-velocity structure (as well as other geophysical data) and the total volume of material encompassed within the Hawaiian Islands. We then used Hawaii as our model for inferring the distribution of volcanic material in various ancient, large igneous provinces. By our calculations, a small province such as Columbia River incorporates approximately 1.3 million cubic kilometers of rock. The giant Ontong Java province contains at least 36 million cubic kilometers of igneous rock, enough to bury the contiguous U.S. under five meters of basalt.

The next stage in understanding the origin and significance of the various igneous provinces is to learn how quickly they formed. Did they accumulate slowly and steadily over tens of millions of years, in much the way that new ocean crust forms at mid-ocean ridges? Or did they emerge from a series of volcanic firestorms that swiftly pumped gases and rock fragments into the air and water and that abruptly transformed the geology of large areas of the earth?

Geologists have expended a great deal of effort attempting to answer those questions for the Deccan Traps. This flood basalt has garnered special attention because it erupted some 65 million years ago, at just about the time of the major extinction at the end of the Cretaceous period, and it may have contributed to that event [see "A Volcanic Eruption," by

Vincent E. Courtillot; *Scientific American,* October 1990]. Robert A. Duncan of Oregon State University, working with Vincent E. Courtillot and Didier Vandamme of the Institute of Physics of the Earth in Paris and several other colleagues, has performed radioactive dating and magnetic analysis on samples from the Deccan Traps. The scientists' results indicate that most of the lavas erupted within a span of less than one million years. Using similar methods, Ajoy K. Baksi of Louisiana State University found that the Columbia River flood basalts mostly erupted over one and a half million years.

Geologists have few direct samples on which to judge the ages and rates of formation of underwater volcanic margins and oceanic plateaus. The scant available evidence hints that submarine igneous provinces accumulated about as rapidly as those on land. Eldholm estimates that the bulk of the North Atlantic volcanic province formed in three million years or less. John A. Tarduno and his collaborators at the Scripps Institution of Oceanography, in conjunction with Mahoney, conclude that the Ontong Java province was constructed in less than three million years. Our analysis of rock dating, conducted by Hugh Davies, formerly of the Australian Geological Survey Organization, and Hubert Whitechurch, formerly of the University of Strasbourg, indicates that the Kerguelen Plateau mostly formed within a span of 4.5 million years.

From a geologic perspective, the largest igneous provinces emerged remarkably fast. In comparison, the Rocky Mountains have been rising for more than 40 million years, and the chain of Hawaiian Islands and the Emperor Seamounts have been building for at least 70 million years. Moreover, much of the volcanic activity associated with the provinces may have occurred in short, violent episodes separated by long periods of relative quiescence.

Once we had collected data on the total volume of the major igneous provinces and on how quickly they formed, we could finally deduce the magnitude of the volcanic forces that created them. The eruptions that built up the Ontong Java province unleashed between 12 and 15 cubic kilometers of igneous rock each year; Deccan volcanism produced between two and eight cubic kilometers annually. Assuming that the creation of igneous provinces, like other eruptive processes, occurs in fits and starts, the pace of crustal production may have been far greater in some years. For reference, Roger L. Larson of the University of Rhode Island estimates that the global network of mid-ocean ridges has yielded between 16 and 26 cubic kilometers of ocean crust a year over the past 150 million years. In other words, individual igneous provinces have generated new crust at

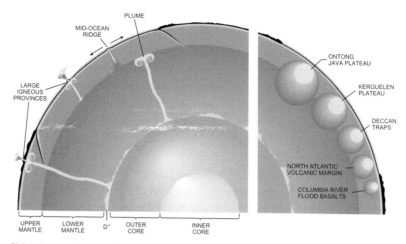

Rising plumes of hot material migrate through the earth's mantle; where the head of the plume reaches the surface, a large igneous province forms (*left*). Plumes probably originate at the boundary layers between the core and mantle (the D″) and between the upper and lower mantle. The parent plumes of the most voluminous igneous provinces were so huge that they must have originated at least in part in the lower mantle, most likely at the D″. The spheres on the right depict the minimum (*white*) and maximum (*gray*) inferred diameters of the plumes associated with five major igneous provinces.

rates comparable to or greater than that of the global seafloor spreading system.

Large igneous provinces build up so swiftly compared with the churning of the earth's deep interior that they must have derived from single, discrete sources. If one knows the volume of basaltic rock contained in those provinces, one can evaluate the dimensions of the hot plumes in the mantle that led to their genesis. Only a fraction of the plume material actually melts and reaches the crust, and this fraction presumably is smaller at greater depths, where increased pressures tend to keep the mantle rocks in a solid state. Hence, less melting should take place under thick continental lithosphere than below thin oceanic lithosphere. Stephen M. Eggins of the Australian National University and Shen-su Sun of the Australian Geological Survey Organization estimate that the molten, basaltic portion of the plume (the part that gives rise to the surface volcanism and to subsurface igneous intrusions) accounts for from 5 to 30 percent of the plume's total volume.

We used those numbers to calculate diameters for the thermal anomalies in the mantle associated with the five best-studied igneous provinces. For simplicity, we assumed the rising plumes to be spherical because such a shape represents the most efficient way to transport material and so permits the plume to move at a plausibly slow velocity through the earth's interior. Our analysis suggests that the Ontong Java province must be derived from a mantle plume at least 600 kilometers, and possibly as much as 1,400 kilometers, in diameter.

That size carries special significance for geologists. It indicates that large plumes must contain at least some material from the lower mantle, more than 670 kilometers below the surface. At that depth, the velocity of seismic waves changes abruptly, probably because of a change in the mineral structure of the mantle rocks. The issue of whether the whole mantle mixes or whether the upper and lower parts of the mantle behave as independent systems that circulate separately strongly divides geophysicists.

Our work favors models that allow for at least some interaction between the upper and lower mantles. In our view, the largest plumes originate in the lower mantle, most likely at the D", a region having unusual seismic properties that lies just outside the core. Smaller plumes may arise at the 670-kilometer-deep transition zone between the upper and lower mantle.

Regardless of where mantle plumes originate, their attributes and effects at the surface depend strongly on the temperature, composition and physical state of the material they encounter just beneath the lithosphere. These factors, combined with the local strength of the lithosphere, determine the volume, timing and position of the eruption at the surface. When plumes rise under continental masses, they may help force the continents apart at a weak spot and then induce the formation of extensive volcanic features along the margins of the rifted landmass [see "Volcanism at Rifts," by Robert S. White and Dan P. McKenzie; *Scientific American*, July 1989]. Under certain circumstances, the plume may penetrate the thick central regions of continental blocks and give birth to a continental flood basalt.

If the upwelling mantle plume surfaces underneath the seafloor, it may give rise to an oceanic plateau. Laboratory experiments suggest that a drawn-out tail of hot material should lag behind the spherical head of the plume, yielding a long-lived, focused source of magma. Over millions of years, plate motions cause the ocean floor to migrate over the site of the hot spot. Where lava erupts at the surface, it gradually constructs a linear submarine ridge or a sequence of islands and seamounts. The Hawaiian-Emperor chain presumably developed in this manner, although how a plume could persist for more than 70 million years remains a puzzle.

As the 1991 eruption of Mount Pinatubo in the Philippines documents, even moderate volcanic outbursts can severely damage the local environ-

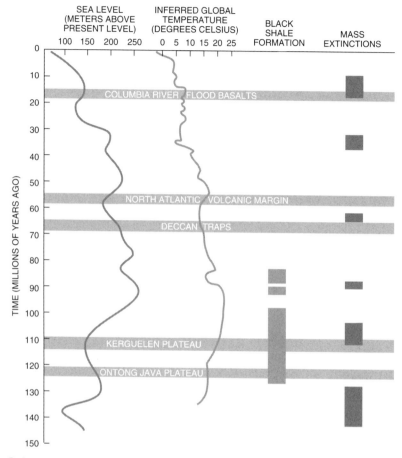

Environmental effects from the eruption of large igneous provinces include fluctuations in the global sea level. Powerful volcanic activity could alter the chemistry and circulation of the atmosphere and oceans, influencing the evolution of life. Volcanoes release carbon dioxide, which may contribute to greenhouse warming; warmer temperatures promote the formation of black shale, in part by enhancing global biomass production. In contrast, several mass extinctions of the past 150 million years coincide with the appearance of igneous provinces, hinting at a causal relationship.

ment. Yet such geologic events are incidental twitches compared with the convulsions of magmatic activity during the formation of large igneous provinces. One can thus begin to imagine that those ancient eruptions had profound consequences. In 1972 Peter R. Vogt of the Naval Research Laboratory first proposed that the surfacing of a mantle plume would lead to physical and chemical changes around the globe; the environmental effects associated with these changes could strongly influence the evolution of life.

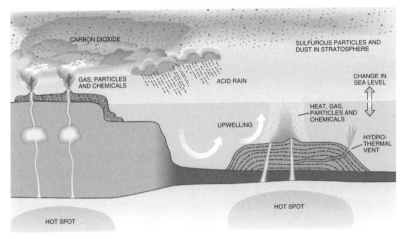

Physical and chemical effects accompany the appearance of large igneous provinces. Volcanoes emit carbon dioxide, which can elevate the global temperature. Sulfurous particles and dust in the stratosphere could induce acid rain and block out sunlight. Heavy metals and other chemicals emitted during eruptions would alter the composition of the land, air and water. Topographic changes associated with extensive volcanism would modify oceanic circulation and change sea level; heat and hydrothermal activity cause upwelling, further transforming underwater conditions.

The current fascination with the greenhouse effect and global change has renewed interest in Vogt's ideas. Stephen Self of the University of Hawaii and Michael R. Rampino of New York University note that the environmental impact of a large igneous province depends in part on whether it forms on land or underwater. Ocean plateaus and volcanic passive margins modify the geometry of the ocean basins and alter the global sea level. We estimate that accumulation of the Ontong Java plateau material elevated sea level by about 10 meters. Oceanic plateaus and volcanic margins may block or constrict circulation in ways that influence large-scale circulation, erosion and sedimentation, especially if the volcanic activity occurs in a sensitive location between ocean basins.

Because they are much denser and more massive than the atmosphere, the oceans are far more able to absorb and dilute gases and heat, a factor that tends to ameliorate the consequences of volcanic eruptions. On the other hand, submarine volcanism and associated hydrothermal activity may leach trace metals such as arsenic, which are poisonous to marine life. Heat produced by submarine eruptions induces ocean-bottom waters to rise to the top, altering the circulation of the surface waters and disrupting the organisms living there.

Carbon dioxide emitted at an undersea volcanic site spreads through the ocean. In so doing, the compound could increase the alkalinity of the seawater, thereby affecting both marine life and climate. If elevated levels of carbon dioxide lead to global warming, ocean circulation should turn sluggish, in which case it would carry less dissolved oxygen. Oxygen-deprived waters may contribute to the formation of carbon-rich black shale, as documented by the late Seymour O. Schlanger of Northwestern University, Michael A. Arthur of Pennsylvania State University and Hugh C. Jenkyns of the University of Oxford.

The atmosphere must have absorbed an extensive outpouring of carbon dioxide around 120 million years ago, when the tempo of volcanism peaked. Larson speculates that a massive thermal instability in the D″ layer initiated so-called superplumes that ultimately supplied the Ontong Java and Kerguelen oceanic plateaus, along with several other smaller plateaus in the Pacific Ocean. Kenneth G. Caldeira of Penn State and Rampino have run computer models of the geochemical cycle based on their assumptions about the quantity of carbon dioxide in the atmosphere during that era. From the models, the workers infer that worldwide temperatures averaged from 7.6 to 12.5 degrees Celsius higher than today's mean, if one takes into account the different geography and higher sea level at the time.

Where volcanic eruptions occur on dry land, they directly alter the physics and chemistry of the atmosphere. Alan R. Huffman of Exxon Exploration Company in Houston calculates that a single flood basalt event that generates 1,000 cubic kilometers of lava (the volume of a typical flow in the Columbia River province) emits 16 trillion kilograms of carbon dioxide, three trillion kilograms of sulfur and 30 billion kilograms of halogens. The thousands of such episodes that must occur in the accumulation of an individual large igneous province would modify the atmosphere in ways that would dwarf the effects of modern, human-generated pollutants.

Explosive eruptions of silica-rich rock often carry sulfurous particles into the stratosphere, where they are converted to tiny droplets of sulfuric acid, according to Rampino and Self. Basaltic lavas emit about 10 times as much sulfur per unit volume as do silica-rich lavas; Charles B. Officer of Dartmouth College and his co-workers deduce that if the gases and particles produced during basaltic eruptions were shot into the stratosphere, they could cause short-term plagues of acid rain, worldwide darkening and global cooling. Richard B. Stothers of the National Aeronautics and Space Administration Goddard Space Flight Center, along with several others, hypothesizes that massive lava fountains and vigorous atmospheric convection taking place over eruptive vents in continental flood basalts could supply the sufficient upward momentum to inject material into the stratosphere.

The powerful Lakagigar eruption in Iceland in 1783–84 illustrates the potential disruptive effects of flood basalt volcanism. Although the Lakagigar eruption poured out only about 15 cubic kilometers of lava, local temperatures declined noticeably in the following years. About three quarters of all livestock in Iceland died, probably the result both of the deterioration of the climate and of the emissions of acid gas; the resulting famine killed about one fourth of Iceland's human population. Dust veils, fog and haze appeared over most of Europe and adjacent parts of Asia and Africa for many months after the eruption.

Scientists are uncovering clear evidence that the environmental consequences of flood basalt volcanism have in fact contributed to mass extinctions. The most severe extinction in terrestrial history occurred 248 million years ago, when the Siberian Traps formed. At that time, roughly 95 percent of all marine species perished; the ensuing evolutionary free-for-all marked the first appearance of the dinosaurs.

The biological repercussions of the eruption of large igneous provinces may depend in part on the state of the global environment at the time. When the environment is already stressed by other factors, such volcanism may trigger rapid climatic, oceanographic and biotic changes. When the environment is robust, few effects may show up in the geologic record.

Oddly enough, the most voluminous igneous province, the Ontong Java, produced virtually no detectable extinctions. On the contrary, its formation coincides with the deposition of black shale, which is suggestive of an interval of enhanced biological activity. We propose that the deleterious effects associated with the Ontong Java were minimal in part because it erupted underwater, as indicated by Ocean Drilling Program studies led by Loren Kroenke of the University of Hawaii and Wolfgang H. Berger of the Scripps Institution of Oceanography. We also suspect that the global environment was in a resilient state 120 million years ago, although such an assertion is difficult to quantify and prove. In contrast, the emplacement of the smaller Kerguelen and Broken Ridge provinces, which took place about 110 million years ago, coincided with a mass extinction as well as with major deposits of black shale.

The eruptions that built up the Deccan Traps, and simultaneous volcanism along the margins of western India and along the Seychelles and the Mascarene Plateau, took place approximately 65 million years ago—just about the time that nearly half of all species of life, including the dinosaurs, went extinct. Scientists continue to debate whether to attribute those changes to the impact of a sizable asteroid or to more earthly explanations. Even if an asteroid was the primary agent of the extinction, the Deccan eruptions may have contributed to an environmental deterioration that could have magnified the repercussions of the impact.

Another significant but less celebrated change in the global environment took place 10 million years after the demise of the dinosaurs, during the emplacement of the North Atlantic volcanic margins. At that time, many deep-ocean foraminifera and land mammals became extinct, and hydrothermal activity was high. Moreover, David K. Rea and his coworkers at the University of Michigan, along with Ellen Thomas of Yale University, find evidence of major transformations in deep water and in atmospheric circulation.

Analysis of the oxygen isotopes taken in by foraminifera indicates that ocean temperatures 55 million years ago, during the early Eocene epoch, were warmer than at any other time during the past 70 million years. Eldholm, working with Thomas, recently postulated that the balmy ocean surface of the early Eocene could have resulted from the outgassing of carbon dioxide during the eruptions along the North Atlantic volcanic margins. Ash layers dating from 55 million years ago cover large regions of northwestern Europe, affirming the violence of the volcanism. Higher atmospheric temperatures could have led to the formation of a layer of warm water atop the oceans at high latitudes. Such a warm surface layer would tend to resist mixing with underlying, cooler waters. The consequent changes in deep-water circulation could have been fatal for many ocean-bottom species.

Even the relatively small Columbia River flood basalts coincided with a mass extinction 16 million years ago. At about that time, the earth began to experience an ongoing cycle of ice ages, as Maureen E. Raymo of the Massachusetts Institute of Technology points out. Perhaps the global environment was already so frail that even a moderate eruption could have had a substantial impact.

Obviously, geologists have—literally and figuratively—only just scratched the surface of large igneous provinces. Current knowledge of these formations amply demonstrates that they contain crucial information about the internal workings of the earth and about the natural causes of global change. Earth scientists are now intensifying their efforts to produce better seismic images, to undertake more field and laboratory studies, to perform additional modeling and to conduct more scientific drilling. Those labors promise even better insight into the intimate links between the earth's inner and outer realms.

FURTHER READING

CONTINENTAL FLOOD BASALTS. Edited by J. D. Macdougall. Kluwer Academic Publishers, 1988.

HOTSPOTS, MANTLE PLUMES, FLOOD BASALTS, AND TRUE POLAR WANDER. R. A. Duncan and M. A. Richards in *Reviews of Geophysics,* Vol. 29, No. 1, pages 31-50; February 1991.

GEOLOGICAL CONSEQUENCES OF SUPERPLUMES. Roger L. Larson in *Geology,* Vol. 19, No. 10, pages 963-966; October 1991.

MAGMATISM AND THE CAUSES OF CONTINENTAL BREAK-UP. Edited by B. C. Storey, T. Alabaster and R. J. Pankhurst. Special Publication, No. 68. Geological Society of London, 1992.

LARGE IGNEOUS PROVINCES: CRUSTAL STRUCTURE, DIMENSIONS, AND EXTERNAL CONSEQUENCES, M. F. Coffin and O. Eldholm in *Reviews of Geophysics.*

Life's Rocky Start

ROBERT M. HAZEN

ORIGINALLY PUBLISHED APRIL 2001

No one knows how life arose on the desolate young earth, but one thing is certain: life's origin was a chemical event. Once the earth formed 4.5 billion years ago, asteroid impacts periodically shattered and sterilized the planet's surface for another half a billion years. And yet, within a few hundred million years of that hellish age, microscopic life appeared in abundance. Sometime in the interim, the first living entity must have been crafted from air, water and rock.

Of those three raw materials, the atmosphere and oceans have long enjoyed the starring roles in origins-of-life scenarios. But rocks, and the minerals of which they are made, have been called on only as bit players or simply as props. Scientists are now realizing that such limited casting is a mistake. Indeed, a recent flurry of fascinating experiments is revealing that minerals play a crucial part in the basic chemical reactions from which life must have arisen.

The first act of life's origin story must have introduced collections of carbon-based molecules that could make copies of themselves. Achieving even this nascent step in evolution entailed a sequence of chemical transformations, each of which added a level of structure and complexity to a group of organic molecules. The most abundant carbon-based compounds available on the ancient earth were gases with only one atom of carbon per molecule, namely, carbon dioxide, carbon monoxide and methane. But the essential building blocks of living organisms—energy-rich sugars, membrane-forming lipids and complex amino acids—may include more than a dozen carbon atoms per molecule. Many of these molecules, in turn, must bond together to form chain-like polymers and other molecular arrays in order to accomplish life's chemical tasks. Linking small molecules into these complex, extended structures must have been especially difficult in the harsh conditions of the early earth, where intense ultraviolet radiation tended to break down clusters of molecules as quickly as they could form.

Carbon-based molecules needed protection and assistance to enact this drama. It turns out that minerals could have served at least five significant functions, from passive props to active players, in life-inducing chemical reactions. Tiny compartments in mineral structures can shelter simple molecules, while mineral surfaces can provide the scaffolding on which those molecules assemble and grow. Beyond these sheltering and supportive functions, crystal faces of certain minerals can actively select particular molecules resembling those that were destined to become biologically important. The metallic ions in other minerals can jump-start meaningful reactions like those that must have converted simple molecules into self-replicating entities. Most surprising, perhaps, are the recent indications that elements of dissolved minerals can be incorporated into biological molecules. In other words, minerals may not have merely helped biological molecules come together, they might have become part of life itself.

PROTECTION FROM THE ELEMENTS

For the better part of a century, following the 1859 publication of Charles Darwin's *On the Origin of Species,* a parade of scientists speculated on life's chemical origins. Some even had the foresight to mention rocks and minerals in their inventive scenarios. But experimental evidence only sporadically buttressed these speculations.

One of the most famous experiments took place at the University of Chicago in 1953. That year chemist Harold C. Urey's precocious graduate student Stanley L. Miller attempted to mimic the earth's primitive oceans and atmosphere in a bottle. Miller enclosed methane, ammonia and other gases thought to be components of the early atmosphere in a glass flask partially filled with water. When he subjected the gas to electric sparks to imitate a prehistoric lightning storm, the clear water turned pink and then brown as it became enriched with amino acids and other essential organic molecules. With this simple yet elegant procedure, Miller transformed origins-of-life research from a speculative philosophical game to an exacting experimental science. The popular press sensationalized the findings by suggesting that synthetic bugs might soon be crawling out of test tubes. The scientific community was more restrained, but many workers sensed that the major obstacle to creating life in the laboratory had been solved.

It did not take long to disabuse researchers of that notion. Miller may have discovered a way to make many of life's building blocks out of the earth's early supply of water and gas, but he had not discovered how or where these simple units would have linked into the complex molecular structures—such as proteins and DNA—that are intrinsic to life.

To answer that riddle, Miller and other origins scientists began proposing rocks as props. They speculated that organic molecules, floating in seawater, might have splashed into tidal pools along rocky coastlines. These molecules would have become increasingly concentrated through repeated cycles of evaporation, like soup thickening in a heated pot.

In recent years, however, researchers have envisioned that life's ingredients might have accumulated in much smaller containers. Some rocks, like gray volcanic pumice, are laced with air pockets created when gases expanded inside the rock while it was still molten. Many common minerals, such as feldspar, develop microscopic pits during weathering. Each tiny chamber in each rock on the early earth could have housed a separate experiment in molecular self-organization. Given enough time and enough chambers, serendipity might have produced a combination of molecules that would eventually deserve to be called "living."

Underlying much of this speculation was the sense that life was so fragile that it depended on rocks for survival. But in 1977 a startling discovery challenged conventional wisdom about life's fragility and, perhaps, its origins. Until then, most scientists had assumed that life spawned at or near the benign ocean surface as a result of chemistry powered by sunlight. That view began to change when deep-ocean explorers first encountered diverse ecosystems thriving at the superheated mouths of volcanic vents on the seafloor. These extreme environments manage to support elaborate communities of living creatures in isolation from the sun. In these dark realms, much of the energy that organisms need comes not from light but from the earth's internal heat. With this knowledge in mind, a few investigators began to wonder whether organic reactions relevant to the origins of life might occur in the intense heat and pressure of these so-called hydrothermal vents.

Miller and his colleagues have objected to the hydrothermal origins hypothesis in part because amino acids decompose rapidly when they are heated. This objection, it turns out, may be applicable only when key minerals are left out of the equation. The idea that minerals might have sheltered the ingredients of life received a boost from recent experiments conducted at my home base, the Carnegie Institution of Washington's Geophysical Laboratory. As a postdoctoral researcher at Carnegie, my colleague Jay A. Brandes (now at the University of Texas Marine Sciences Institute in Port Aransas) proposed that minerals help delicate amino acids remain intact. In 1998 we conducted an experiment in which the amino acid leucine broke down within a matter of minutes in pressurized water at 200 degrees Celsius—just as Miller and his colleagues predicted. But when Brandes added to the mix an iron sulfide mineral of the type commonly found in and around hydrothermal vents, the amino acid stayed intact for days—plenty of time to react with other critical molecules.

A ROCK TO STAND ON

Even if the right raw materials were contained in a protected place—whether it was a tidal pool, a microscopic pit in a mineral surface or somewhere inside the plumbing of a seafloor vent—the individual molecules would still be suspended in water. These stray molecules needed a support structure—some kind of scaffolding—where they could cling and react with one another.

One easy way to assemble molecules from a dilute solution is to concentrate them on a flat surface. Errant molecules might have been drawn to the calm surface of a tidal pool or perhaps to a primitive "oil slick" of compounds trapped at the water's surface. But such environments would have posed a potentially fatal hazard to delicate molecules. Harsh lightning storms and ultraviolet radiation accosted the young earth in doses many times greater than they do today. Such conditions would have quickly broken the bonds of complex chains of molecules.

Origins scientists with a penchant for geology have long recognized that minerals might provide attractive alternative surfaces where important molecules could assemble. Like the container idea, this notion was born half a century ago. At that time, a few scientists had begun to suspect that clays have special abilities to attract organic molecules. These ubiquitous minerals feel slick when wet because their atoms form flat, smooth layers. The surfaces of these layers frequently carry an electric charge, which might be able to attract organic molecules and hold them in place. Experiments later confirmed these speculations. In the late 1970s an Israeli research group demonstrated that amino acids can concentrate on clay surfaces and then link up into short chains that resemble biological proteins. These chemical reactions occurred when the investigators evaporated a water-based solution containing amino acids from a vessel containing clays—a situation not unlike the evaporation of a shallow pond or tidal pool with a muddy bottom.

More recently, separate research teams led by James P. Ferris of the Rensselaer Polytechnic Institute and by Gustaf Arrhenius of the Scripps Institution of Oceanography demonstrated that clays and other layered minerals can attract and assemble a variety of organic molecules. In a tour de force series of experiments during the past decade, the team at Rensselaer found that clays can act as scaffolds for the building blocks of RNA, the molecule in living organisms that translates genetic instructions into proteins.

Once organic molecules had attached themselves to a mineral scaffold, various types of complex molecules could have been forged. But only a chosen few were eventually incorporated into living cells. That means that some kind of template must have selected the primitive molecules that would become biologically important. Recent experiments show, once again, that minerals may have played a central role in this task.

PREFERENTIAL TREATMENT

Perhaps the most mysterious episode of selection left all living organisms with a strange predominance of one type of amino acid. Like many organic molecules, amino acids come in two forms. Each version comprises the same types of atoms, but the two molecules are constructed as mirror images of each other. The phenomenon is called chirality, but for simplicity's sake scientists refer to the two versions as "left-handed" (or "L") and "right-handed" (or "D"). Organic synthesis experiments like Miller's invariably produce 50–50 mixtures of L and D molecules, but the excess of left-handed amino acids in living organisms is nearly 100 percent.

Researchers have proposed a dozen theories—from the mundane to the exotic—to account for this bizarre occurrence. Some astrophysicists have argued that the earth might have formed with an excess of L amino acids—a consequence of processes that took place in the cloud of dust and gas that became the solar system. The main problem with this theory is that in most situations such processes yield only the slightest excess—less than 1 percent—of L or D molecules.

Alternatively, the world might have started with a 50–50 mixture of L and D amino acids, and then some important feature of the physical environment selected one version over the other. To me, the most obvious candidates for this specialized physical environment are crystal faces whose surface structures are mirror images of each other. Last spring I narrowed in on calcite, the common mineral that forms limestone and marble, in part because it often displays magnificent pairs of mirror-image faces. The chemical structure of calcite in many mollusk shells bonds strongly to amino acids. Knowing this, I began to suspect that calcite surfaces may feature chemical bonding sites that are ideally suited to only one type of amino acid or the other. With the help of my Carnegie colleague Timothy

Filley (now at Purdue University) and Glenn Goodfriend of George Washington University, I ran more than 100 tests of this hypothesis.

Our experiments were simple in concept, although they required meticulous clean-room procedures to avoid contamination by the amino acids that exist everywhere in the environment. We immersed a well-formed, fist-size crystal of calcite into a 50–50 solution of aspartic acid, a common amino acid. After 24 hours we removed the crystal from this solution, washed it in water and carefully collected all the molecules that had adhered to specific crystal faces. In one experiment after another we observed that calcite's "left-handed" faces selected L-amino acids, and vice versa, with excesses approaching 40 percent in some cases.

Curiously, calcite faces with finely terraced surfaces displayed the greatest selectivity. This outcome led us to speculate that these terraced edges might force the L and D amino acids to line up in neat rows on their respective faces. Under the right environmental conditions, these organized rows of amino acids might chemically join to form proteinlike molecules—some made entirely of L amino acids, others entirely of D. If protein formation can indeed occur, this result becomes even more exciting, because recent experiments by other investigators indicate that some proteins can self-replicate. In the earth's early history, perhaps a self-replicating protein formed on the face of a calcite crystal.

Left- and right-handed crystal faces occur in roughly equal numbers, so chiral selection of L amino acids probably did not happen everywhere in the world at once. Our results and predictions instead suggest that the first successful set of self-replicating molecules—the precursor to all the varied life-forms on the earth today—arose at a specific time and place. It was purely chance that the successful molecule developed on a crystal face that preferentially selected left-handed amino acids over their right-handed counterparts.

Minerals undoubtedly could have acted as containers, scaffolds and templates that helped to select and organize the molecular menagerie of the primitive earth. But many of us in origins research suspect that minerals played much more active roles, catalyzing key synthesis steps that boosted the earth's early inventory of complex biological molecules.

GETTING A JUMP ON THE ACTION

Experiments led by Carnegie researcher Brandes in 1997 illustrate this idea. Biological reactions require nitrogen in the form of ammonia, but the only common nitrogen compound thought to have been available on the primitive earth is nitrogen gas. Perhaps, Brandes thought, the environment at hydrothermal vents mimics an industrial process in which ammonia is synthesized by passing nitrogen and hydrogen over a hot metallic surface. Sure enough, when we subjected hydrogen, nitrogen and the iron oxide mineral magnetite to the pressures and temperatures characteristic of a sea-floor vent, the mineral catalyzed the synthesis of ammonia.

The idea that minerals may have triggered life's first crucial steps has emerged most forcefully from the landmark theory of chemist Günter Wächtershäuser, a German patent lawyer with a deep interest in life's origins. In 1988 Wächtershäuser advanced a sweeping theory of organic evolution in which minerals—mostly iron and nickel sulfides that abound at deep-sea hydrothermal vents—could have served as the template, the catalyst and the energy source that drove the formation of biological molecules. Indeed, he has argued that primitive living entities were molecular coatings that adhered to the positively charged surfaces of pyrite, a mineral composed of iron and sulfur. These entities, he further suggests, obtained energy from the chemical reactions that produce pyrite. This hypothesis makes sense in part because some metabolic enzymes—the molecules that help living cells process energy—have at their core a cluster of metal and sulfur atoms.

For much of the past three years, Wächtershäuser's provocative theory has influenced our experiments at Carnegie. Our team, including geochemist George Cody and petrologist Hatten S. Yoder, has focused on the possibility that metabolism can proceed without enzymes in the presence of minerals—especially oxides and sulfides. Our simple strategy, much in the spirit of Miller's famous experiment, has been to subject ingredients known to be available on the young earth—water, carbon dioxide and minerals—to a controlled environment. In our case, we try to replicate the bone-crushing pressures and scalding temperatures typical of a deep-sea hydrothermal vent. Most of our experiments test the interactions among ingredients enclosed in welded gold capsules, which are roughly the size of a daily vitamin pill. We place as many as six capsules into Yoder's "bomb"—a massive steel pressure chamber that squeezes the tiny capsules to pressures approaching 2,000 atmospheres and heats them to about 250 degrees C.

One of our primary goals in these organic-synthesis experiments—and one of life's fundamental chemical reactions—is carbon fixation, the process of producing molecules with an increasing number of carbon atoms in their chemical structure. Such reactions follow two different paths depending on the mineral we use. We find that many common minerals, including most oxides and sulfides of iron, copper and zinc, promote carbon addition by a routine industrial process known as Fischer-Tropsch (F-T) synthesis.

This process can build chainlike organic molecules from carbon monoxide and hydrogen. First, carbon monoxide and hydrogen react to form methane, which has one carbon atom. Adding more carbon monoxide and hydrogen to the methane produces ethane, a two-carbon molecule, and then the reaction repeats itself, adding a carbon atom each time. In the chemical industry, researchers have harnessed this reaction to manufacture molecules with virtually any desired number of carbon atoms. Our first organic-synthesis experiments in 1996, and much more extensive research by Thomas McCollom of the Woods Hole Oceanographic Institution, demonstrate that F-T reactions can build molecules with 30 or more carbon atoms under some hydrothermal-vent conditions in less than a day. If this process manufactures large organic molecules from simple inorganic chemicals throughout the earth's hydrothermal zones today, then it very likely did so in the planet's prebiological past.

When we conduct experiments using nickel or cobalt sulfides, we see that carbon addition occurs primarily by carbonylation—the insertion of a carbon and oxygen molecule, or carbonyl group. Carbonyl groups readily attach themselves to nickel or cobalt atoms, but not so strongly that they cannot link to other molecules and jump ship to form larger molecules. In one series of experiments, we observed the lengthening of the nine-carbon molecule nonyl thiol to form 10-carbon decanoic acid, a compound similar to the acids that drive metabolic reactions in living cells. What is more, all the reactants in this experiment—a thiol, carbon monoxide and water—are readily available near sulfide-rich hydrothermal vents. By repeating these simple kinds of reactions—adding a carbonyl group here or a hydroxide group there—we can synthesize a rich variety of complex organic molecules.

Our 1,500 hydrothermal organic synthesis experiments at Carnegie have done more than supplement the catalogue of interesting molecules that must have been produced on the early earth. These efforts reveal another, more complex behavior of minerals that may have significant consequences for the chemistry of life. Most previous origins-of-life studies have treated minerals as solid and unchanging—stable platforms where organic molecules could assemble. But we are finding that in the presence of hot water at high pressure, minerals start to dissolve. In the process, the liberated atoms and molecules from the minerals can become crucial reactants in the primordial soup.

THE HEART OF THE MATTER

Our first discovery of minerals as reactants was an unexpected result of our recent catalysis experiments led by Cody. As expected, carbonylation reactions produced 10-carbon decanoic acid from a mixture of simple molecules inside our gold capsules. But significant quantities of elemental sulfur, organic sulfides, methyl thiol and other sulfur compounds appeared as well. The sulfur in all these products must have been liberated from the iron sulfide mineral.

Even more striking was the liberation of iron, which brilliantly colored the water-based solutions inside the capsules. As the mineral dissolved, the iron formed bright red and orange organometallic complexes in which iron atoms are surrounded by various organic molecules. We are now investigating the extent to which these potentially reactive complexes might act as enzymes that promote the synthesis of molecular structures.

The role of minerals as essential chemical ingredients of life is not entirely unexpected. Hydrothermal fluids are well known to dissolve and concentrate mineral matter. At deep-sea vents, spectacular pillars of sulfide grow dozens of feet tall as plumes of hot, mineral-laden water rise from below the seafloor, contact the frigid water of the deep ocean and deposit new layers of minerals on the growing pillar. But the role of these dissolved minerals has not yet figured significantly in origins scenarios. Whatever their behavior, dissolved minerals seem to make the story of life's emergence much more interesting.

When we look beyond the specifics of prebiological chemistry, it is clear that the origin of life was far too complex to imagine as a single event. Rather we must work from the assumption that it was a gradual sequence of more modest events, each of which added a degree of order and complexity to the world of prebiological molecules. The first step must have been the synthesis of the basic building blocks. Half a century of research reveals that the molecules of life were manufactured in abundance—in the nebula that formed our solar system, at the ocean's surface, and near hydrothermal vents. The ancient earth suffered an embarrassment of riches—a far greater diversity of molecules than life could possibly employ.

Minerals helped to impose order on this chaos. First by confining and concentrating molecules, then by selecting and arranging those molecules, minerals may have jump-started the first self-replicating molecular systems. Such a system would not have constituted life as we know it, but it could have, for the first time, displayed a key property of life. In this scenario, a self-replicating molecular system began to use up the resources of its environment. As mutations led to slightly different variants, competition for limited resources initiated and drove the process of molecular natural selection. Self-replicating molecular systems began to evolve, inevitably becoming more efficient and more complex.

A long-term objective for our work at the Carnegie Institution is to demonstrate simple chemical steps that could lead to a self-replicating system—perhaps one related to the metabolic cycles common to all living cells. Scientists are far from creating life in the laboratory, and it may never be possible to prove exactly what chemical transformations gave rise to life on earth. What we can say for sure is that minerals played a much more complex and integral part in the origin of life than most scientists ever suspected. By being willing to cast minerals in starring roles in experiments that address life's beginnings, researchers may come closer to answering one of science's oldest questions.

FURTHER READING

BEGINNINGS OF CELLULAR LIFE. Harold J. Morowitz. Yale University Press, 1992.

ORIGINS OF LIFE: THE CENTRAL CONCEPTS. David W. Deamer and Gail R. Fleischaker. Jones and Bartlett, 1994.

EMERGENCE: FROM CHAOS TO ORDER. John H. Holland. Helix Books, 1998.

BIOGENESIS: THEORIES OF LIFE'S ORIGIN. Noam Lahav. Oxford University Press, 1999.

When Methane Made Climate

JAMES F. KASTING

ORIGINALLY PUBLISHED JULY 2004

About 2.3 billion years ago unusual microbes breathed new life into young Planet Earth by filling its skies with oxygen. Without those prolific organisms, called cyanobacteria, most of the life that we see around us would never have evolved.

Now many scientists think another group of single-celled microbes were making the planet habitable long before that time. In this view, oxygen-detesting methanogens reigned supreme during the first two billion years of Earth's history, and the greenhouse effect of the methane they produced had profound consequences for climate.

Scientists first began to suspect methane's dramatic role more than 20 years ago, but only during the past four years have the various pieces of the ancient methane story come together. Computer simulations now reveal that the gas—which survives about 10 years in today's atmosphere—could have endured for as long as 10,000 years in an oxygen-free world. No fossil remains exist from that time, but many microbiologists believe that methanogens were some of the first life-forms to evolve. In their prime, these microbes could have generated methane in quantities large enough to stave off a global deep freeze. The sun was considerably dimmer then, so the added greenhouse influence of methane could have been exactly what the planet needed to keep warm. But the methanogens did not dominate forever. The plummeting temperatures associated with their fading glory could explain Earth's first global ice age and perhaps others as well.

The prevalence of methane also means that a pinkish-orange haze may have shrouded the planet, as it does Saturn's largest moon, Titan. Although Titan's methane almost certainly comes from a nonbiological source, that moon's similarities to the early Earth could help reveal how greenhouse gases regulated climate in our planet's distant past.

FAINT SUN FOILED

When Earth formed some 4.6 billion years ago, the sun burned only 70 percent as brightly as it does today [see "How Climate Evolved on the Terrestrial Planets," by James F. Kasting, Owen B. Toon and James B. Pollack; *Scientific American,* February 1988]. Yet the geologic record contains no convincing evidence for widespread glaciation until about 2.3 billion years ago, which means that the planet was probably even warmer than during the modern cycle of ice ages of the past 100,000 years. Thus, not only did greenhouse gases have to make up for a fainter sun, they also had to maintain average temperatures considerably higher than today's.

Methane was far from scientists' first choice as an explanation of how the young Earth avoided a deep freeze. Because ammonia is a much stronger greenhouse gas than methane, Carl Sagan and George H. Mullen of Cornell University suggested in the early 1970s that it was the culprit. But later research showed that the sun's ultraviolet rays rapidly destroy ammonia in an oxygen-free atmosphere. So this explanation did not work.

Another obvious candidate was carbon dioxide (CO_2), one of the primary gases spewing from the volcanoes abundant at that time. Although they debated the details, most scientists assumed for more than 20 years that this gas played the dominant role. In 1995, however, Harvard University researchers uncovered evidence that convinced many people that CO_2 levels were too low to have kept the early Earth warm.

The Harvard team, led by Rob Rye, knew from previous studies that if the atmospheric concentrations of CO_2 had exceeded about eight times the present-day value of around 380 parts per million (ppm), the mineral siderite ($FeCO_3$) would have formed in the top layers of the soil as iron reacted with CO_2 in the oxygen-free air. But when the investigators studied samples of ancient soils from between 2.8 billion and 2.2 billion years ago, they found no trace of siderite. Its absence implied that the CO_2 concentration must have been far less than would have been needed to keep the planet's surface from freezing.

Even before CO_2 lost top billing as the key greenhouse gas, researchers had begun to explore an alternative explanation. By the late 1980s, scientists had learned that methane traps more heat than an equivalent concentration of CO_2 because it absorbs a wider range of wavelengths of Earth's outgoing radiation. But those early studies underestimated methane's influence. My group at Pennsylvania State University turned to methane because we knew that it would have had a much longer lifetime in the ancient atmosphere.

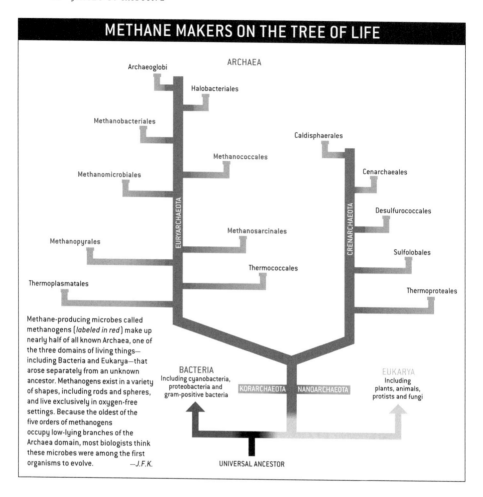

METHANE MAKERS ON THE TREE OF LIFE

ARCHAEA

Archaeoglobi

Halobacteriales

Methanobacteriales

Caldisphaerales

Methanococcales

Cenarchaeales

Methanomicrobiales

Desulfurococcales

EURYARCHAEOTA

CRENARCHAEOTA

Methanosarcinales

Methanopyrales

Sulfolobales

Thermococcales

Thermoplasmatales

Thermoproteales

Methane-producing microbes called methanogens (*labeled in red*) make up nearly half of all known Archaea, one of the three domains of living things—including Bacteria and Eukarya—that arose separately from an unknown ancestor. Methanogens exist in a variety of shapes, including rods and spheres, and live exclusively in oxygen-free settings. Because the oldest of the five orders of methanogens occupy low-lying branches of the Archaea domain, most biologists think these microbes were among the first organisms to evolve.   —J.F.K.

BACTERIA
Including cyanobacteria, proteobacteria and gram-positive bacteria

KORARCHAEOTA   NANOARCHAEOTA

EUKARYA
Including plants, animals, protists and fungi

UNIVERSAL ANCESTOR

In today's oxygen-rich atmosphere, the carbon in methane is much happier teaming up with the oxygen in hydroxyl radicals to produce CO_2 and carbon monoxide (CO), releasing water vapor in the process. Consequently, methane remains in the atmosphere a mere 10 years and plays just a bit part in warming the planet. Indeed, the gas exists in minuscule concentrations of only about 1.7 ppm; CO_2 is roughly 220 times as concentrated at the planet's surface and water vapor 6,000 times.

To determine how much higher those methane concentrations must have been to warm the early Earth, my students and I collaborated with researchers from the NASA Ames Research Center to simulate the ancient climate. When we assumed that the sun was 80 percent as bright as today, which is the value expected 2.8 billion years ago, an atmosphere with no methane at all would have had to contain a whopping 20,000 ppm of CO_2 to keep the surface temperature above freezing. That concentration is 50 times as high as modern values and seven times as high as the upper limit on CO_2 that the studies of ancient soils revealed. When the simulations calculated CO_2 at its maximum possible value, the atmosphere required the help of 1,000 ppm of methane to keep the mean surface temperature above freezing—in other words, 0.1 percent of the atmosphere needed to be methane.

UP TO THE TASK?

The early atmosphere could have maintained such high concentrations only if methane was being produced at rates comparable to today. Were primordial methanogens up to the task? My colleagues and I teamed up with microbiologist Janet L. Siefert of Rice University to try to find out.

Biologists have several reasons to suspect that such high methane levels could have been maintained. Siefert and others think that methane-producing microbes were some of the first microorganisms to evolve. They also suggest that methanogens would have filled niches that oxygen producers and sulfate reducers now occupy, giving them a much more prominent biological and climatic role than they have in the modern world.

Methanogens would have thrived in an environment fueled by volcanic eruptions. Many methanogens feed directly on hydrogen gas (H_2) and CO_2 and belch methane as a waste product; others consume acetate and other compounds that form as organic matter decays in the absence of oxygen. That is why today's methanogens can live only in oxygen-free environments such as the stomachs of cows and the mud under flooded rice paddies. On the early Earth, however, the entire atmosphere was devoid of oxygen, and volcanoes released significant amounts of H_2. With no oxygen available to form water, H_2 probably accumulated in the atmosphere and oceans in concentrations high enough for methanogens to use.

Based on these and other considerations, some scientists have proposed that methanogens living on geologically derived hydrogen might form the base of underground microbial ecosystems on Mars and on Jupiter's ice-covered moon, Europa. Indeed, a recent report from the European Space Agency's Mars Express spacecraft suggests that the present Martian atmosphere may contain approximately 10 parts per billion of methane. If verified, this finding would be consistent with having methanogens living below the surface of Mars.

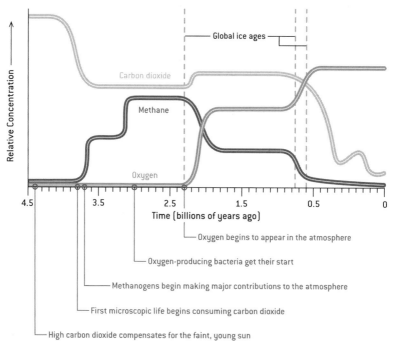

Relative concentrations of major atmospheric gases may explain why global ice ages [*dashed lines*] occurred in Earth's distant past. Methane-producing microorganisms flourished initially, but as oxygen skyrocketed about 2.3 billion years ago, these microbes suddenly found few environments where they could survive. The accompanying decrease in methane—a potent greenhouse gas—could have chilled the entire planet. The role of carbon dioxide, the most notable greenhouse gas in today's atmosphere, was probably much less dramatic.

Geochemists estimate that on the early Earth H_2 reached concentrations of hundreds to thousands of parts per million—that is, until methanogens evolved and converted most of it to methane. Thermodynamic calculations reveal that if other essential nutrients, such as phosphorus and nitrogen, were available, methanogens would have used most of the available H_2 to make methane. (Most scientists agree that sufficient phosphorus would have come from the chemical breakdown of rocks and that various ocean-dwelling microorganisms were producing plenty of nitrogen.) In this scenario, the methanogens would have yielded the roughly 1,000 ppm of methane called for by the computer models to keep the planet warm.

Even more evidence for the primordial dominance of methanogens surfaced when microbiologists considered how today's methanogens would have reacted to a steamy climate. Most methanogens grow best at temperatures above 40 degrees Celsius; some even prefer at least 85 degrees C. Those that thrive at higher temperatures grow faster, so as the intensifying greenhouse effect raised temperatures at the planet's surface, more of these faster-growing, heat-loving specialists would have survived. As they made up a larger proportion of the methanogen population, more methane molecules would have accumulated in the atmosphere, making the surface temperature still warmer—in fact, hotter than today's climate, despite the dimmer sun.

SMOG SAVES THE DAY

As a result of that positive feedback loop, the world could have eventually become such a hothouse that life itself would have been difficult for all but the most extreme heat–loving microbes. This upward spiral could not have continued indefinitely, however. Once atmospheric methane becomes more abundant than CO_2, methane's reaction to sunlight changes. Instead of being oxidized to CO or CO_2, it polymerizes, or links together, to form complex hydrocarbons that then condense into particles, forming an organic haze. Planetary scientists observe a similar haze in the atmosphere of Saturn's largest moon: Titan's atmosphere consists primarily of molecular nitrogen, N_2, along with a small percentage of methane. The scientists hope to learn more when NASA's Cassini spacecraft arrives at Saturn in July [see "Saturn at Last!" by Jonathan I. Lunine; *Scientific American,* June 2004].

The possible formation of organic haze in Earth's young atmosphere adds a new wrinkle to the climate story. Because they form at high altitudes, these particles have the opposite effect on climate that greenhouse gases do. A greenhouse gas allows most visible solar radiation to pass through, but it absorbs and reradiates outgoing infrared radiation, thereby warming the surface. In contrast, high-altitude organic haze absorbs incoming sunlight and reradiates it back into space, thereby reducing the total amount of radiation that reaches the surface. On Titan, this so-called antigreenhouse effect cools the surface by seven degrees C or so. A similar haze layer on the ancient Earth would have also cooled the climate, thus shifting the methanogen population back toward those slower-growing species that prefer cooler weather and thereby limiting further increases in methane production. This powerful negative feedback loop would have tended to stabilize Earth's temperature and atmospheric composition at exactly the point at which the layer of organic haze began to form.

NOTHING LASTS FOREVER

Methane-induced smog kept the young Earth comfortably warm—but not forever. Global ice ages occurred at least three times in the period known as the Proterozoic eon, first at 2.3 billion years ago and again at 750 million and 600 million years ago. The circumstances surrounding these glaciations were long unexplained, but the methane hypothesis provides compelling answers here as well.

The first of these glacial periods is often called the Huronian glaciation because it is well exposed in rocks just north of Lake Huron in southern Canada. Like the better-studied late Proterozoic glaciations, the Huronian event appears to have been global, based on interpretations that some of the continents were near the equator at the time ice covered them.

This cold snap formed layers of jumbled rocks and other materials that a glacier carried and then dropped to the ground when the ice melted sometime between 2.45 billion and 2.2 billion years ago. In the older rocks below these glacial deposits are detrital uraninite and pyrite, two minerals considered evidence for very low levels of atmospheric oxygen. Above the glacial layers sits a red sandstone containing hematite—a mineral that forms only under oxygen-rich skies. (Hematite has also been found at the landing site of the Mars rover Opportunity. This hematite is gray, however, because the grain size is larger.) The layering of these distinct rock types indicates that the Huronian glaciations occurred precisely when atmospheric oxygen levels first rose.

This apparent coincidence remained unexplained until recently; if we hypothesize that methane kept the ancient climate warm, then we can predict a global ice age at 2.3 billion years ago because it would have been a natural consequence of the rise of oxygen. Many of the methanogens and other anaerobic organisms that dominated the planet before the rise of oxygen would have either perished in this revolution or found themselves confined to increasingly restricted habitats.

Although this finale sounds as if it is the end of the methane story, that is not necessarily the case. Methane never again exerted a dominating effect on climate, but it could still have been an important influence at later times—during the late Proterozoic, for example, when some scientists suggest that the oceans froze over entirely during a series of so-called snowball Earth episodes [see "Snowball Earth," by Paul F. Hoffman and Daniel P. Schrag; *Scientific American*, January 2000].

Indeed, methane concentrations could have remained significantly higher than today's during much of the Proterozoic eon, which ended about 600 million years ago, if atmospheric oxygen had continued to be somewhat lower and the deep oceans were still anoxic and low in sulfate, a dissolved salt common in modern seawater. The rate at which methane escaped from the seas to the atmosphere could still have been up to 10 times as high as it is now, and the concentration of methane in the atmosphere could have been as high as 100 ppm. This scenario might explain why the Proterozoic remained ice-free for almost a billion and a half years despite the fact that the sun was still relatively dim. My colleagues and I have speculated that a second rise in atmospheric oxygen, or in dissolved sulfate, could conceivably have triggered the snowball Earth episodes as well—once again by decreasing the warming presence of methane.

EXTRATERRESTRIAL METHANE

As compelling as this story of methanogens once ruling the world may sound, scientists are forced to be content with no direct evidence to back it up. Finding a rock that contains bubbles of ancient atmosphere would provide absolute proof, but such a revelation is unlikely. The best we can say is that the hypothesis is consistent with several indirect pieces of evidence—most notably, the low atmospheric CO_2 levels inferred from ancient soils and the timing of the first planet-encompassing ice age.

Although we may never be able to verify this hypothesis on Earth, we may be able to test it indirectly by observing Earth-like planets orbiting other stars. Both NASA and the European Space Agency are designing large space-based telescopes to search for Earth-size planets orbiting some 120 nearby stars. If such planets exist, these missions—NASA's Terrestrial Planet Finder and ESA's Darwin—should be able to scan their atmospheres for the presence of gases that would indicate the existence of life.

Oxygen at any appreciable abundance would almost certainly indicate biology comparable to that of modern Earth, provided that the planet was also endowed with the liquid water necessary for life. High levels of methane, too, would suggest some form of life. As far as we know, on planets with Earth-like surface temperatures only living organisms can produce methane at high levels. The latter discovery might be one of the best ways for scientists to gain a better understanding of what our own planet was like during the nascent stages of its history.

FURTHER READING

GREENHOUSE WARMING BY CH$_4$ IN THE ATMOSPHERE OF EARLY EARTH. Alexander A. Pavlov, James F. Kasting, Lisa L. Brown, Kathy A. Rages and Richard Freedman in *Journal of Geophysical Research—Planets*, Vol. 105, No. E5, pages 11,981–11,990; May 2000.

LIFE AND THE EVOLUTION OF EARTH'S ATMOSPHERE. James F. Kasting and Janet L. Siefert in *Science*, Vol. 296, pages 1066–1068; May 10, 2002.

METHANE-RICH PROTEROZOIC ATMOSPHERE? Alexander A. Pavlov, Matthew T. Hurtgen, James F. Kasting and Michael A. Arthur in *Geology*, Vol. 31, No. 1, pages 87–90; January 2003.

Snowball Earth

PAUL F. HOFFMAN AND DANIEL P. SCHRAG

ORIGINALLY PUBLISHED IN JANUARY 2000

Our human ancestors had it rough. Saber-toothed cats and woolly mammoths may have been day-to-day concerns, but harsh climate was a consuming long-term challenge. During the past million years, they faced one ice age after another. At the height of the last icy episode, 20,000 years ago, glaciers more than two kilometers thick gripped much of North America and Europe. The chill delivered ice as far south as New York City.

Dramatic as it may seem, this extreme climate change pales in comparison to the catastrophic events that some of our earliest microscopic ancestors endured around 600 million years ago. Just before the appearance of recognizable animal life, in a time period known as the Neoproterozoic, an ice age prevailed with such intensity that even the tropics froze over.

Imagine the earth hurtling through space like a cosmic snowball for 10 million years or more. Heat escaping from the molten core prevents the oceans from freezing to the bottom, but ice grows a kilometer thick in the −50 degree Celsius cold. All but a tiny fraction of the planet's primitive organisms die. Aside from grinding glaciers and groaning sea ice, the only stir comes from a smattering of volcanoes forcing their hot heads above the frigid surface. Although it seems the planet might never wake from its cryogenic slumber, the volcanoes slowly manufacture an escape from the chill: carbon dioxide.

With the chemical cycles that normally consume carbon dioxide halted by the frost, the gas accumulates to record levels. The heat-trapping capacity of carbon dioxide—a greenhouse gas—warms the planet and begins to melt the ice. The thaw takes only a few hundred years, but a new problem arises in the meantime: a brutal greenhouse effect. Any creatures that survived the icehouse must now endure a hothouse.

As improbable as it may sound, we see clear evidence that this striking climate reversal—the most extreme imaginable on this planet—happened as many as four times between 750 million and 580 million years ago. Scientists long presumed that the earth's climate was never so severe; such intense climate change has been more widely accepted for other planets such as Venus [see "Global Climate Change on Venus," by Mark A. Bullock and David H. Grinspoon; *Scientific American*, March 1999]. Hints of a harsh past on the earth began cropping up in the early 1960s, but we and our colleagues have found new evidence in the past eight years that has helped us weave a more explicit tale that is capturing the attention of geologists, biologists and climatologists alike.

Thick layers of ancient rock hold the only clues to the climate of the Neoproterozoic. For decades, many of those clues appeared rife with contradiction. The first paradox was the occurrence of glacial debris near sea level in the tropics. Glaciers near the equator today survive only at 5,000 meters above sea level or higher, and at the worst of the last ice age they reached no lower than 4,000 meters. Mixed in with the glacial debris are unusual deposits of iron-rich rock. These deposits should have been able to form only if the Neoproterozoic oceans and atmosphere contained little or no oxygen, but by that time the atmosphere had already evolved to nearly the same mixture of gases as it has today. To confound matters, rocks known to form in warm water seem to have accumulated just after the glaciers receded. If the earth were ever cold enough to ice over completely, how did it warm up again? In addition, the carbon isotopic signature in the rocks hinted at a prolonged drop in biological productivity. What could have caused this dramatic loss of life?

Each of these long-standing enigmas suddenly makes sense when we look at them as key plot developments in the tale of a "snowball earth." The theory has garnered cautious support in the scientific community since we first introduced the idea in the journal *Science* a year and a half ago. If we turn out to be right, the tale does more than explain the mysteries of Neoproterozoic climate and challenge long-held assumptions about the limits of global change. These extreme glaciations occurred just before a rapid diversification of multicellular life, culminating in the so-called Cambrian explosion between 575 and 525 million years ago. Ironically, the long periods of isolation and extreme environments on a snowball earth would most likely have spurred on genetic change and could help account for this evolutionary burst.

The search for the surprisingly strong evidence for these climatic events has taken us around the world. Although we are now examining Neoproterozoic rocks in Australia, China, the western U.S. and the Arctic islands of Svalbard, we began our investigations in 1992 along the rocky

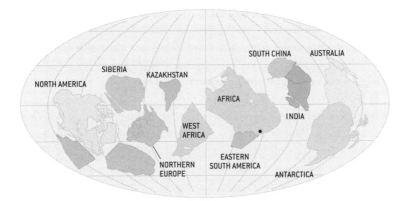

Earth's landmasses were most likely clustered near the equator during the global glaciations that took place around 600 million years ago. Although the continents have since shifted position, relics of the debris left behind when the ice melted are exposed at dozens of points on the present land surface, including what is now Namibia (*dot*).

cliffs of Namibia's Skeleton Coast. In Neoproterozoic times, this region of southwestern Africa was part of a vast, gently subsiding continental shelf located in low southern latitudes.

There we see evidence of glaciers in rocks formed from deposits of dirt and debris left behind when the ice melted. Rocks dominated by calcium- and magnesium-carbonate minerals lie just above the glacial debris and harbor the chemical evidence of the hothouse that followed. After hundreds of millions of years of burial, these now exposed rocks tell the story that scientists first began to piece together 35 years ago.

In 1964 W. Brian Harland of the University of Cambridge pointed out that glacial deposits dot Neoproterozoic rock outcrops across virtually every continent. By the early 1960s scientists had begun to accept the idea of plate tectonics, which describes how the planet's thin, rocky skin is broken into giant pieces that move atop a churning mass of hotter rock below. Harland suspected that the continents had clustered together near the equator in the Neoproterozoic, based on the magnetic orientation of tiny mineral grains in the glacial rocks. Before the rocks hardened, these grains aligned themselves with the magnetic field and dipped only slightly relative to horizontal because of their position near the equator. (If they had formed near the poles, their magnetic orientation would be nearly vertical.)

Realizing that the glaciers must have covered the tropics, Harland became the first geologist to suggest that the earth had experienced a great Neoproterozoic ice age [see "The Great Infra-Cambrian Glaciation," by W. B. Harland and M.J.S. Rudwick; *Scientific American,* August 1964]. Although some of Harland's contemporaries were skeptical about the reliability of the magnetic data, other scientists have since shown that Harland's hunch was correct. But no one was able to find an explanation for how glaciers could have survived the tropical heat.

At the time Harland was announcing his ideas about Neoproterozoic glaciers, physicists were developing the first mathematical models of the earth's climate. Mikhail Budyko of the Leningrad Geophysical Observatory found a way to explain tropical glaciers using equations that describe the way solar radiation interacts with the earth's surface and atmosphere to control climate. Some geographic surfaces reflect more of the sun's incoming energy than others, a quantifiable characteristic known as albedo. White snow reflects the most solar energy and has a high albedo, darker-colored seawater has a low albedo, and land surfaces have intermediate values that depend on the types and distribution of vegetation.

The more radiation the planet reflects, the cooler the temperature. With their high albedo, snow and ice cool the atmosphere and thus stabilize their own existence. Budyko knew that this phenomenon, called the ice-albedo feedback, helps modern polar ice sheets to grow. But his climate simulations also revealed that this feedback can run out of control. When ice formed at latitudes lower than around 30 degrees north or south of the equator, the planet's albedo began to rise at a faster rate because direct sunlight was striking a larger surface area of ice per degree of latitude. The feedback became so strong in his simulation that surface temperatures plummeted and the entire planet froze over.

FROZEN AND FRIED

Budyko's simulation ignited interest in the fledgling science of climate modeling, but even he did not believe the earth could have actually experienced a runaway freeze. Almost everyone assumed that such a catastrophe would have extinguished all life, and yet signs of microscopic algae in rocks up to one billion years old closely resemble modern forms and imply a continuity of life. Also, once the earth had entered a deep freeze, the high albedo of its icy veneer would have driven surface temperatures so low that it seemed there would have been no means of escape. Had such a glaciation occurred, Budyko and others reasoned, it would have been permanent.

The first of these objections began to fade in the late 1970s with the discovery of remarkable communities of organisms living in places once thought too harsh to harbor life. Seafloor hot springs support microbes that thrive on chemicals rather than sunlight. The kind of volcanic activity that feeds the hot springs would have continued unabated in a snowball earth. Survival prospects seem even rosier for psychrophilic, or cold-loving, organisms of the kind living today in the intensely cold and dry mountain valleys of East Antarctica. Cyanobacteria and certain kinds of algae occupy habitats such as snow, porous rock and the surfaces of dust particles encased in floating ice.

The key to the second problem—reversing the runaway freeze—is carbon dioxide. In a span as short as a human lifetime, the amount of carbon dioxide in the atmosphere can change as plants consume the gas for photosynthesis and as animals breathe it out during respiration. Moreover, human activities such as burning fossil fuels have rapidly loaded the air with carbon dioxide since the beginning of the Industrial Revolution in the late 1700s. In the earth's lifetime, however, these carbon sources and sinks become irrelevant compared with geologic processes.

Carbon dioxide is one of several gases emitted from volcanoes. Normally this endless supply of carbon is offset by the erosion of silicate rocks: The chemical breakdown of the rocks converts carbon dioxide to bicarbonate, which is washed to the oceans. There bicarbonate combines with calcium and magnesium ions to produce carbonate sediments, which store a great deal of carbon [see "Modeling the Geochemical Carbon Cycle," by R. A. Berner and A. C. Lasaga; *Scientific American,* March 1989].

In 1992 Joseph L. Kirschvink, a geobiologist at the California Institute of Technology, pointed out that during a global glaciation, an event he termed a snowball earth, shifting tectonic plates would continue to build volcanoes and to supply the atmosphere with carbon dioxide. At the same time, the liquid water needed to erode rocks and bury the carbon would be trapped in ice. With nowhere to go, carbon dioxide would collect to incredibly high levels—high enough, Kirschvink proposed, to heat the planet and end the global freeze.

Kirschvink had originally promoted the idea of a Neoproterozoic deep freeze in part because of mysterious iron deposits found mixed with the glacial debris. These rare deposits are found much earlier in earth history when the oceans (and atmosphere) contained very little oxygen and iron could readily dissolve. (Iron is virtually insoluble in the presence of oxygen.) Kirschvink reasoned that millions of years of ice cover would deprive the oceans of oxygen, so that dissolved iron expelled from seafloor hot springs could accumulate in the water. Once a carbon dioxide–induced greenhouse effect began melting the ice, oxygen would again mix with the seawater and force the iron to precipitate out with the debris once carried by the sea ice and glaciers.

With this greenhouse scenario in mind, climate modelers Kenneth Caldeira of Lawrence Livermore National Laboratory and James F. Kasting of Pennsylvania State University estimated in 1992 that overcoming the runaway freeze would require roughly 350 times the present-day concentration of carbon dioxide. Assuming volcanoes of the Neoproterozoic belched out gases at the same rate as they do today, the planet would have remained locked in ice for up to tens of millions of years before enough carbon dioxide could accumulate to begin melting the sea ice. A snowball earth would be not only the most severe conceivable ice age, it would be the most prolonged.

CARBONATE CLUES

Kirschvink was unaware of two emerging lines of evidence that would strongly support his snowball earth hypothesis. The first is that the Neo-

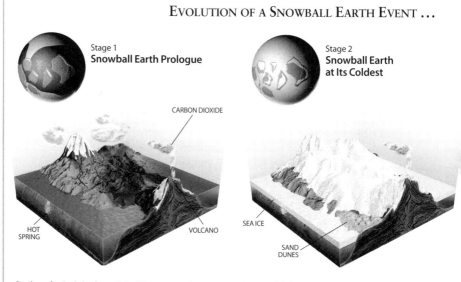

EVOLUTION OF A SNOWBALL EARTH EVENT ...

Stage 1
Snowball Earth Prologue

Stage 2
Snowball Earth at Its Coldest

CARBON DIOXIDE

HOT SPRING

VOLCANO

SEA ICE

SAND DUNES

Breakup of a single landmass 770 million years ago leaves small continents scattered near the equator. Formerly landlocked areas are now closer to oceanic sources of moisture. Increased rainfall scrubs more heat-trapping carbon dioxide out of the air and erodes continental rocks more quickly. Consequently, global temperatures fall, and large ice packs form in the polar oceans. The white ice reflects more solar energy than does darker seawater, driving temperatures even lower. This feedback cycle triggers an unstoppable cooling effect that will engulf the planet in ice within a millennium.

Average global temperatures plummet to –50 degrees Celsius shortly after the runaway freeze begins. The oceans ice over to an average depth of more than a kilometer, limited only by heat emanating slowly from the earth's interior. Most microscopic marine organisms die, but a few cling to life around volcanic hot springs. The cold, dry air arrests the growth of land glaciers, creating vast deserts of windblown sand. With no rainfall, carbon dioxide emitted from volcanoes is not removed from the atmosphere. As carbon dioxide accumulates, the planet warms and sea ice slowly thins.

proterozoic glacial deposits are almost everywhere blanketed by carbonate rocks. Such rocks typically form in warm, shallow seas, such as the Bahama Banks in what is now the Atlantic Ocean. If the ice and warm water had occurred millions of years apart, no one would have been surprised. But the transition from glacial deposits to these "cap" carbonates is abrupt and lacks evidence that significant time passed between when the glaciers dropped their last loads and when the carbonates formed. Geologists were stumped to explain so sudden a change from glacial to tropical climates.

Pondering our field observations from Namibia, we realized that this change is no paradox. Thick sequences of carbonate rocks are the expected consequence of the extreme greenhouse conditions unique to the transient aftermath of a snowball earth. If the earth froze over, an ultrahigh carbon dioxide atmosphere would be needed to raise temperatures to the melting point at the equator. Once melting begins, low-albedo seawater replaces high-albedo ice and the runaway freeze is reversed [*see illustration below*].

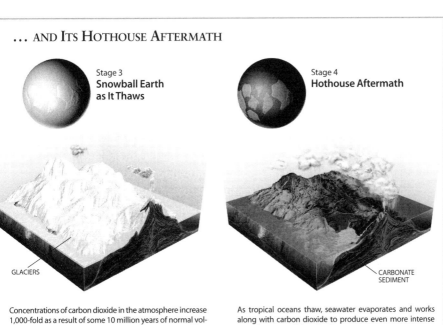

... AND ITS HOTHOUSE AFTERMATH

Stage 3
**Snowball Earth
as It Thaws**

Stage 4
Hothouse Aftermath

GLACIERS

CARBONATE
SEDIMENT

Concentrations of carbon dioxide in the atmosphere increase 1,000-fold as a result of some 10 million years of normal volcanic activity. The ongoing greenhouse warming effect pushes temperatures to the melting point at the equator. As the planet heats up, moisture from sea ice sublimating near the equator refreezes at higher elevations and feeds the growth of land glaciers. The open water that eventually forms in the tropics absorbs more solar energy and initiates a faster rise in global temperatures. In a matter of centuries, a brutally hot, wet world will supplant the deep freeze.

As tropical oceans thaw, seawater evaporates and works along with carbon dioxide to produce even more intense greenhouse conditions. Surface temperatures soar to more than 50 degrees Celsius, driving an intense cycle of evaporation and rainfall. Torrents of carbonic acid rain erode the rock debris left in the wake of the retreating glaciers. Swollen rivers wash bicarbonate and other ions into the oceans, where they form carbonate sediment. New life-forms—engendered by prolonged genetic isolation and selective pressure—populate the world as global climate returns to normal.

The greenhouse atmosphere helps to drive surface temperatures upward to almost 50 degrees C, according to calculations made last summer by climate modeler Raymond T. Pierrehumbert of the University of Chicago.

Resumed evaporation also helps to warm the atmosphere because water vapor is a powerful greenhouse gas, and a swollen reservoir of moisture in the atmosphere would drive an enhanced water cycle. Torrential rain would scrub some of the carbon dioxide out the air in the form of carbonic acid, which would rapidly erode the rock debris left bare as the glaciers subsided. Chemical erosion products would quickly build up in the ocean water, leading to the precipitation of carbonate sediment that would rapidly accumulate on the seafloor and later become rock. Structures preserved in the Namibian cap carbonates indicate that they accumulated extremely rapidly, perhaps in only a few thousand years. For example, crystals of the mineral aragonite, clusters of which are as tall as a person, could precipitate only from seawater highly saturated in calcium carbonate.

Cap carbonates harbor a second line of evidence that supports Kirschvink's snowball escape scenario. They contain an unusual pattern in the ratio of two isotopes of carbon: common carbon 12 and rare carbon 13, which has an extra neutron in its nucleus. The same patterns are observed in cap carbonates worldwide, but no one thought to interpret them in terms of a snowball earth. Along with Alan Jay Kaufman, an isotope geochemist now at the University of Maryland, and Harvard University graduate student Galen Pippa Halverson, we have discovered that the isotopic variation is consistent over many hundreds of kilometers of exposed rock in northern Namibia.

Carbon dioxide moving into the oceans from volcanoes is about 1 percent carbon 13; the rest is carbon 12. If the formation of carbonate rocks were the only process removing carbon from the oceans, then the rock would have the same fraction of carbon 13 as that which comes out of volcanoes. But the soft tissues of algae and bacteria growing in seawater also use carbon from the water around them, and their photosynthetic machinery prefers carbon 12 to carbon 13. Consequently, the carbon that is left to build carbonate rocks in a life-filled ocean such as we have today has a higher ratio of carbon 13 to carbon 12 than does the carbon fresh out of a volcano.

The carbon isotopes in the Neoproterozoic rocks of Namibia record a different situation. Just before the glacial deposits, the amount of carbon 13 plummets to levels equivalent to the volcanic source, a drop we think records decreasing biological productivity as ice encrusted the oceans at high latitudes and the earth teetered on the edge of a runaway freeze. Once the oceans iced over completely, productivity would have essentially ceased, but no carbon record of this time interval exists because calcium carbonate could not have formed in an ice-covered ocean. This drop in carbon 13 persists through the cap carbonates atop the glacial deposits and then gradually rebounds to higher levels of carbon 13 several hundred meters above, presumably recording the recovery of life at the end of the hothouse period.

Abrupt variation in this carbon isotope record shows up in carbonate rocks that represent other times of mass extinction, but none are as large or as long-lived. Even the meteorite impact that killed off the dinosaurs 65 million years ago did not bring about such a prolonged collapse in biological activity.

Overall, the snowball earth hypothesis explains many extraordinary observations in the geologic record of the Neoproterozoic world: the carbon isotopic variations associated with the glacial deposits, the paradox of cap carbonates, the evidence for long-lived glaciers at sea level in the tropics, and the associated iron deposits. The strength of the hypothesis is that it simultaneously explains all these salient features, none of which had satisfactory independent explanations. What is more, we believe this hypothesis sheds light on the early evolution of animal life.

SURVIVAL AND REDEMPTION OF LIFE

In the 1960s Martin J. S. Rudwick, working with Brian Harland, proposed that the climate recovery following a huge Neoproterozoic glaciation paved the way for the explosive radiation of multicellular animal life soon thereafter. Eukaryotes—cells that have a membrane-bound nucleus and from which all plants and animals descended—had emerged more than one billion years earlier, but the most complex organisms that had evolved when the first Neoproterozoic glaciation hit were filamentous algae and unicellular protozoa. It has always been a mystery why it took so long for these primitive organisms to diversify into the 11 animal body plans that show up suddenly in the fossil record during the Cambrian explosion [*see illustration on next page*].

A series of global freeze-fry events would have imposed an environmental filter on the evolution of life. All extant eukaryotes would thus stem from the survivors of the Neoproterozoic calamity. Some measure of the extent of eukaryotic extinctions may be evident in universal "trees of life." Phylogenetic trees indicate how various groups of organisms evolved

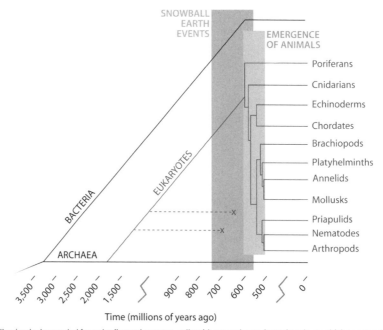

Time (millions of years ago)

All animals descended from the first eukaryotes, cells with a membrane-bound nucleus, which appeared about two billion years ago. By the time of the first snowball earth episode more than one billion years later, eukaryotes had not developed beyond unicellular protozoa and filamentous algae. But despite the extreme climate, which may have "pruned" the eukaryote tree (*dashed lines*), all 11 animal phyla ever to inhabit the earth emerged within a narrow window of time in the aftermath of the last snowball event. The prolonged genetic isolation and selective pressure intrinsic to a snowball earth could be responsible for this explosion of new life-forms.

from one another, based on their degrees of similarity. These days biologists commonly draw these trees by looking at the sequences of nucleic acids in living organisms.

Most such trees depict the eukaryotes' phylogeny as a delayed radiation crowning a long, unbranched stem. The lack of early branching could mean that most eukaryotic lineages were "pruned" during the snowball earth episodes. The creatures that survived the glacial episodes may have taken refuge at hot springs both on the seafloor and near the surface of the ice where photosynthesis could be maintained.

The steep and variable temperature and chemical gradients endemic to ephemeral hot springs would preselect for survival in the hellish aftermath to come. In the face of varying environmental stress, many organisms respond with wholesale genetic alterations. Severe stress encourages a great degree of genetic change in a short time, because organisms that can most quickly alter their genes will have the most opportunities to acquire traits that will help them adapt and proliferate.

Hot-spring communities widely separated geographically on the icy surface of the globe would accumulate genetic diversity over millions of years. When two groups that start off the same are isolated from each other long enough under different conditions, chances are that at some point the extent of genetic mutation will produce a new species. Repopulations occurring after each glaciation would come about under unusual and rapidly changing selective pressures quite different from those preceding the glaciation; such conditions would also favor the emergence of new life-forms.

Martin Rudwick may not have gone far enough with his inference that climatic amelioration following the great Neoproterozoic ice age paved the way for early animal evolution. The extreme climatic events themselves may have played an active role in spawning multicellular animal life.

We have shown how the worldwide glacial deposits and carbonate rocks in the Neoproterozoic record point to an extraordinary type of climatic event, a snowball earth followed by a briefer but equally noxious greenhouse world. But what caused these calamities in the first place, and why has the world been spared such events in more recent history? The first possibility to consider is that the Neoproterozoic sun was weaker by approximately 6 percent, making the earth more susceptible to a global freeze. The slow warming of our sun as it ages might explain why no snowball event has occurred since that time. But convincing geologic evidence suggests that no such glaciations occurred in the billion or so years before the Neoproterozoic, when the sun was even cooler.

The unusual configuration of continents near the equator during Neoproterozoic times may better explain how snowball events get rolling [*see illustration on page 47*]. When the continents are nearer the poles, as they are today, carbon dioxide in the atmosphere remains in high enough concentrations to keep the planet warm. When global temperatures drop enough that glaciers cover the high-latitude continents, as they do in Antarctica and Greenland, the ice sheets prevent chemical erosion of the rocks beneath the ice. With the carbon burial process stifled, the carbon dioxide in the atmosphere stabilizes at a level high enough to fend off the advancing ice sheets. If all the continents cluster in the tropics, on the other hand, they would remain ice-free even as the earth grew colder and approached the critical threshold for a runaway freeze. The carbon dioxide "safety switch" would fail because carbon burial continues unchecked.

We may never know the true trigger for a snowball earth, as we have but simple theories for the ultimate forcing of climate change, even in recent times. But we should be wary of the planet's capacity for extreme change. For the past million years, the earth has been in its coldest state since animals first appeared, but even the greatest advance of glaciers 20,000 years ago was far from the critical threshold needed to plunge the earth into a snowball state. Certainly during the next several hundred years, we will be more concerned with humanity's effects on climate as the earth heats up in response to carbon dioxide emissions [see "The Human Impact on Climate Change," by Thomas R. Karl and Kevin E. Trenberth; *Scientific American,* December 1999]. But could a frozen world be in our more distant future?

We are still some 80,000 years from the peak of the next ice age, so our first chance for an answer is far in the future. It is difficult to say where the earth's climate will drift over millions of years. If the trend of the past million years continues and if the polar continental safety switch were to fail, we may once again experience a global ice catastrophe that would inevitably jolt life in some new direction.

FURTHER READING

ORIGIN AND EARLY EVOLUTION OF THE METAZOA. Edited by J. H. Lipps and P. W. Signor. Plenum Publishing, 1992.

THE ORIGIN OF ANIMAL BODY PLANS. D. Erwin, J. Valentine and D. Jablonski in *American Scientist,* Vol. 85, No. 2, pages 126–137; March–April 1997.

A NEOPROTEROZOIC SNOWBALL EARTH. P. F. Hoffman, A. J. Kaufman, G. P. Halverson and D. P. Schrag in *Science,* Vol. 281, pages 1342–1346; August 28, 1998.

THE FIRST ICE AGE. Kristin Leutwyler. Available only at www.sciam.com/2000/0100issue/0100hoffman.html on the *Scientific American* Web site.

Looking for Life below the Bottom

SARAH SIMPSON

ORIGINALLY PUBLISHED IN JUNE 2000

Paul Johnson squints at a video monitor that's glowing with a live image of the barren seafloor nearly three kilometers beneath the white caps of the North Pacific. The monitor is his porthole through the single eye of Jason, a submersible robot hovering at the end of an electrical and fiber-optic tether that dangles from the research vessel *Thomas G. Thompson.*

Arms folded across his chest, Johnson glances at the engineer who pilots the robot from his seat in the shipboard control room. The crew has been mentally puttering about the ocean bottom for 13 hours straight, but Johnson doesn't want to miss the task ahead. They have only 30 minutes to snake the six-foot hose of a pump into an opening in the seafloor before an automatic timer switches the pump on. In the dim video image, Johnson sees the water above the hole shimmering with heat. This water is rising from the rocks *underneath* the ocean, and Johnson believes that something in the water is alive—or once was.

The pilot takes a breath and a sip of coffee, then twists a joystick on the panel of gadgets before him. The robot extends its single aluminum arm and four-fingered claw toward the bare hose, which is hardly thicker than the robot's half-inch-wide finger. Grabbing the hose is easier said than done, because using the robot handicaps the pilot in several ways: It offers him no depth perception to gauge distance and no sense of touch to check his grip. To glance left or right, he must push buttons that tilt the camera. Every move happens in slow motion. From the shadows, Johnson and a band of observers grimace when the white hose slithers out of reach. "It's like watching a video game when the other kid won't let you play," Johnson says. Time ticks forward as the pilot flies the robot closer to the evasive hose—and the shimmering water.

Finding microscopic creatures rising from a seafloor hot spring wouldn't surprise anyone if that spring were a so-called black smoker. Isolated from the sun's life-giving energy, those towering rock chimneys spew scalding fountains of chemicals that give rise to entire food chains of creatures, from sulfur-eating bacteria on up to red-tipped tube worms and predatory spider crabs. Black smokers are fascinating but rare. All of those explored so far add up to no more than a few acres of seafloor.

This warm spring is different: it emanates from a man-made hole, and it is 100 kilometers from the nearest-known black smoker. Johnson and fellow marine scientist Jim Cowen suspect that if suboceanic life is gushing out here, then an inconspicuous Eden of immense proportions very likely dwells inside the fractured rock that underlies all the oceans. This subsurface realm could sustain more living stuff than all marine habitats combined.

It is difficult to resist pondering the global implications of such a discovery. Fed by minerals from within and warmed by magma from below, were fledgling life-forms hiding in the ocean crust while the young Earth endured asteroid assaults and global deep freezes? And if life thrives within Earth's rocky foundation, then why not under the oceans of Jupiter's ice-covered moon Europa or in the riverbeds of ancient Mars?

Only 11 minutes now remain on the pump timer. Johnson knows that the pump is the quickest and cleanest way to capture signs of life from the crust. If it clicks on in normal seawater, the sample will be worthless. The robot finally grasps the hose firmly, but the pilot struggles to maneuver the hose into the hole. He just misses, and the audience gasps. At last, the hose enters the shimmering water with a minute to spare.

Tense and frustrating to watch, this operation is not so different from what it might be like to conduct a robotic search for life on Europa or Mars. But whether in space or at sea, small victories sometimes create unexpected problems, as the scientists on board the *Thompson* are soon reminded. Squeezing the hose for three hours overheats Jason's fingers, and the crippled robot must temporarily cease its work.

Indeed, frustration at sea was what first drew Johnson and Cowen together, in 1994. That fall the two scientists, then acquaintances conducting unrelated research, ended up on the same ship. "We had crummy weather, we had equipment failures, we had operator errors—we had a lot of spare time," Johnson remembers. "So Jim and I started talking." They soon discovered that they shared a controversial idea—that life might flourish in the ocean under the seafloor—and thought they saw a way to test it. Their peers reacted skeptically, but after Johnson and Cowen twice revised their approach, the National Science Foundation came through with $1.6 million.

But a single research program might not be enough. Even if the microbial wonderland that they hypothesize does exist, it may take decades to prove it. In most places the crust lies beneath a few kilometers of seawater—not to mention a mound of muck deep enough to bury a 60-story skyscraper. The two scientists knew that the best chance to prove their theory was to find spots in the seafloor where heat from below forces water out of the rock.

On their first voyage, in 1997, they located a few eligible leaks. A year later they capped these leaks with sophisticated filters designed to snare cells, DNA or other signs of life that are flushed out with the water. In late August 1999 they set sail once more. On this 13-day expedition, they need Jason and its nine-man crew from the Woods Hole Oceanographic Institution in Massachusetts to retrieve the six filters they planted on the seafloor in 1998. They also want to test a new pump and a smaller, cleaner filter that Cowen's team designed to stay down for only a week.

Departing from its home port at the University of Washington in Seattle, the *Thompson* followed a westerly course toward a jagged gash in the seafloor called the Juan de Fuca Ridge, which mirrors the coastline some 350 kilometers from land. Part of a worldwide web of lava-belching cracks, the ridge is forested with black smokers, appropriately named: Godzilla, Inferno, Hulk. But Johnson and Cowen are not investigating those hot spots. The *Thompson* instead stopped 100 kilometers short of the ridge, above the relatively uncharted Cascadia Basin, a low spot between the coast and the ridge. Unwinding a giant spool of steel-armored cable, the Woods Hole engineers slowly lowered Jason toward an ordinary patch of cold seabed, where molten lava stiffened into rock called basalt more than three million years ago. Cold rock hardly seems a tantalizing haven for throngs of microscopic life, but Johnson and Cowen see things differently.

For one thing, the old crust isn't as solid as you might think. Johnson, a 59-year-old geophysicist from the University of Washington, has spent much of his career getting to know ocean basalt. By studying minuscule variations in the gravitational field of the seafloor, he has estimated that up to 20 percent of the top 600 meters of the rock in this area is composed of water-filled voids. "Every single place where there's water, space and temperatures below 100 degrees Celsius, you've got life," Johnson states, reciting the mantra of his fellow believers. (Then again, this confidence comes from a man who told a student in 1967 that plate tectonics is a ridiculous idea. "So I have a long tradition of being absolutely wrong," he jokes.)

Certainly many biologists will think that Johnson is absolutely wrong about suboceanic life, too, unless he and Cowen collect irrefutable evidence. Their biggest challenge is to capture unique life from below without allowing any common life from above to sneak into the samples. That may sound like a simple task, but microbes are notoriously opportunistic. They can scrape together a living from simple molecules of carbon, sulfur and iron, which are found almost everywhere: in seawater, in mud overlying the rocks—even on the metallic surfaces of scientific devices. Accessing the deep crust with a drill only magnifies the risk of contamination. Like pulling a stopper in a bathtub, punching a hole in the ocean bottom usually allows the life-infested ocean to rush in.

Not so with the hole now framed in Jason's video camera. There warm water flows from the top of a steel straw that extends through 250 meters of mud and penetrates 50 meters into the rock below. Pressure has been driving water out of that hole since it was drilled three years ago and, perhaps, has flushed out contamination. When Cowen first heard about this drill hole, the 48-year-old biologist says he felt "like a candy-crazed kid being given the keys to a candy store." He and his group from the University of Hawaii are attempting to gather suboceanic life inside a four-foot section of plumbing pipe connected to the steel straw. Inside the plastic column is a filter that has been scavenging signs of life from the warm water flowing through it for the past year. Or so they hope.

The Jason pilot maneuvers the robot for a closer look at a small titanium cylinder riding piggyback on the scavenging column. The cylinder is loaded with electronic sensors that measure water temperature and flow rates, which help Johnson and his students to see whether the conditions could possibly support life. A scientist frowns as he notices the green fuzz of corrosion around brass connectors in the titanium lid—so much corrosion that he wonders if the device will make it to the surface without leaking. High-pressure leaks could trigger an explosion at the surface. Even a small blast would be bad, because the scavenging column contains mercuric chloride, a deadly poison that the biologists use to kill the crustal critters trapped in the filter and to keep the seawater bugs away.

"The poison was meant to percolate through slowly to kill things over time," Cowen explains as the team transfers the scavenging column from the seafloor to the ship's deck for a cleansing shower. "I think it's long gone." Back in the biology laboratory, he and his students break into the column with no problems. Cowen changes his blue surgical gloves, careful to be clean before reaching inside for the filter. Delicately but briskly, he pulls out the folded layers of woven glass fibers, now saturated with greenish-brown slime. "These were pristine white when we put them in here," he comments. A smile spreads over his face as he and his students divvy segments of the filter into protective glass tubes.

The green slime is just what they had hoped to find. They collected a similar but smaller sample here last year and gave it to Steve Giovannoni, a molecular biologist at Oregon State University known for his expertise in deciphering the genetic code of seawater microbes. "We know what's in seawater, and this stuff looks different," Giovannoni told them before they left. But "different" isn't proof that the bugs came from the crust. Cowen knows that the more organic debris they collect, the better their chances to rule out contamination. That's why they want to send down the new pump and filter before they move to the next site.

Four days into the cruise the problem that began with Jason's overheated claw mushrooms into a crisis that halts the team's ambitious agenda. "There's no joy in Mudville today," Johnson remarks as he paces among the cranes, winch and unused buoys crowding the ship's aft deck. A dozen albatross gather off the stern. Nature is siding with science on this calm, sun-drenched afternoon, but the machines are not. Once again the engineers are reeling Jason back to the ship. The robot had nearly reached the bottom when they discovered that its claw, supposedly repaired, was still paralyzed.

Johnson, who has lived some two years of his life at sea, expects this kind of trouble in deep-ocean research. "At least half of what goes on are disasters," he says. Yet he claims to never grow weary. His childhood dream was to go to the moon, but polio left him with a limp. "So," he explains, "I went the other way." That decision took place a quarter of a century ago. Veterans like Johnson know that their most agonizing moments at sea will be selecting which carefully planned activities to abandon. They intentionally overbook each expedition so as not to waste precious ship time, which costs about $20,000 per day.

By sunset the engineers have amputated Jason's malfunctioning arm. A black trash bag protects the vulnerable stump from salt spray, and the mineral oil that keeps the robot's insides from imploding at depth dribbles from a Plexiglas chamber. Inside the workshop one engineer rebuilds the claw while two others rewire the electronics. As always, the team works through the night, but it fails to diagnose Jason's affliction.

Johnson and the other leaders decide to reshuffle the cruise timetable to give the Jason crew more time. Two days ahead of schedule the captain plots a course to Axial volcano, a blister of young crust along the ridge.

There they expect to find a more diverse and populous neighborhood of microbes, which will help them imagine the subsea milieus that must exist between here and there. At lunch the next day someone asks Johnson what he plans to do when they arrive. "If the arm is working, we'll dive and do a lot of things; if the arm isn't working, we'll dive and do a great deal less," Johnson replies in his typical even-keeled manner. The truth is, without Jason's arm they have no hope of retrieving the five scavenging columns still sitting on the seafloor.

The 12-hour journey to Axial gives the engineers the time they need. As soon as the *Thompson* reaches the underwater volcano, the sub descends within a few meters of pillowlike mounds of glassy basalt, evidence of molten rock solidifying on contact with frigid seawater. Through white webs of bacteria, floating like sheets of disintegrating tissue paper, the scientists spot a column perched on the platform that they had cemented to the bare rock last year.

They haul the column on board, and Johnson checks the water-flow sensor. He is fascinated by an intriguing pattern: the volume of water coming out of the seabed fluctuates in sync with the passing tides. It is well known that the ocean over the Juan de Fuca ridge changes depth by two or three meters with each tidal cycle. What Johnson has discovered is that the weight of the passing tidal bulge compresses the rock, which expands again when the tidal bulge moves on. Heat alone moves water through the rock at the rate of only a few meters per year. "That's really slow," Johnson points out. "But if you take that same environment and squeeze it every 12 hours, then you've got velocities on the order of meters per *day*." Microbes living in the rock, which crave chemical nutrients dissolved in the circulating water, couldn't ask for a quicker and more reliable delivery service.

The team picks up a third column, also in young crust, then turns back toward the Cascadia Basin. The *Thompson* arrives at the drill hole with only a day to spare before it must return to shore. Early that afternoon Jason grabs the small column they put down a week ago, and Cowen's crew hauls it to the lab to see what is stuck in its newfangled filter. The ship has already begun to move six kilometers south to an elevated spot of crust that barely protrudes above the muck. There, where warm water rises in hot curtains from a maze of cracks in the rock, await the last three scavenging columns left behind when the ship departed early for Axial.

Cowen snaps on a fresh pair of blue gloves and loads swing music into the lab stereo. Jason has been performing without a hiccup since the dives at Axial. Best of all, the team has accomplished most of its highest priorities—at this point, it has recovered half of the scavenging columns, all of which performed as planned during an entire year on the ocean bottom. The final tasks are all but complete: Jason has just packed two of the remaining columns in a buoyant basket and is ascending with the third hooked to its frame.

The buzz of the drill Cowen uses to open up the small column drowns out the brassy blare of the horns and the beat of the drums. He beams as he pulls out the round, sandwichlike filter, now clogged with the familiar green ooze. "This is as clean as it gets," Cowen notes as he slices the filter like a miniature pie.

Johnson enters the room, looking serious. The basket seems to be holding the last two columns hostage on the seafloor. It is not responding to commands to drop its weights, and the crew worries it won't surface before nightfall, only five hours away. (In an arrangement similar to that of a hot-air balloon, aluminum bars connect the basket to a team of floats, which are impossible to spot in the dark despite their bright-yellow color.) The task can't wait until morning, because the barometer is dropping and the waves are mounting after a two-week calm. In a brief but intense discussion, Johnson, Cowen, the ship's captain and the Jason crew leader decide to send the robot down to release the weights. During the three hours it takes for the basket to surface, the weather deteriorates.

The swells are now surging to nine feet. Johnson stands at the rail wearing a hard hat, life vest and rubber gloves—he has volunteered to close the safety valves on the columns' poison boxes before the crew brings the basket on deck. The captain nudges the heaving, 3,200-ton bulk of the *Thompson* within spitting distance of the bobbing yellow floats. Stretching over the rail and flailing with a metal hook in hand, the chief mate snags a rope tied to the basket's frame and hooks it to a crane. Now lifted just above the water, the floats can no longer hold the basket upright, and an ominous gray swell tilts it sideways.

Johnson sees that one of the two scavenging columns was loaded heavy-side-up and now protrudes precariously. The crane operator throws the winch to full power as the passing wave allows the basket to hang free for an instant. The next swell again flips it on its side, and the inverted column teeters on the container's edge. A third swell engulfs the basket, and after the longest second of the cruise, it pops above the spray with *neither* column inside. Johnson lunges to the railing to watch both instruments sink slowly out of sight.

Losing $25,000 in equipment is bad, but losing the year's worth of clues caught in the two filters is unthinkably upsetting. The chief mate launches into apologies, but Johnson assures him that none are necessary.

He turns toward the crowd lining the rail of the deck above him—a collage of faces looking as stunned as he feels. Cowen is among them, but the two scientists don't bother speaking until later that evening, when the ship is headed to port. "There was no villainy, no incompetence," Johnson will later say, "just the wrong wave at the wrong time." They can only hope that the filters they did retrieve hold the answers they are looking for. At this point, they have no money to come back and try again.

Several months later Cowen receives an e-mail report from an organic chemist in Chicago who has been studying some of the samples from the Cascadia Basin. It turns out that the water the team collected at the drill hole contains fatty cell fragments, called lipids, of a kind that doesn't show up in the mud or seawater samples from the same site. Giovannoni has other exciting news: he has discovered in the once shimmering water a diverse array of microbes, among them heat-loving bacteria that could never survive the icy temperatures of the ocean bottom and anaerobic bacteria that would choke on the oxygen present in normal seawater. "The names tell the story," Giovannoni says. "Show this list to any microbiologists, and they'd say it was a jackpot." But this jackpot is still short of proof that the bugs came from the crust. The same slew of microbes could also live happily in other warm, oxygen-deprived locales—namely, the steel pipe that lines the drill hole.

Many more months of arduous laboratory analyses may turn up something more definitive. In the meantime, Johnson and some of his colleagues are asking the National Science Foundation to finance a return trip to the site of the lost columns. His new team wants to drive a 10-foot titanium pipe into the bare rock to serve as a permanent spigot for tapping fluids—and a potential multitude of microscopic creatures—in the crust. "Our titanium hypodermic needle would be giving Mother Earth a type of blood test," Johnson says. "Only in this case, we would be quite pleased if she came up 'infected.'"

MARINE LIFE

Life in the Ocean

JAMES W. NYBAKKEN AND STEVEN K. WEBSTER

ORIGINALLY PUBLISHED IN *SCIENTIFIC AMERICAN PRESENTS*, FALL 1998

Earth is misnamed. Even though the planet is largely made up of rock, 71 percent of its surface is covered with ocean. Like the wet film coating a newly washed plum, this water makes up a thin layer compared with the globe as a whole. Yet that watery veneer comprises more than 90 percent of the biosphere by volume: it covers 360 million square kilometers (140 million square miles) and runs, on average, a few kilometers deep. It is in these salty depths that life first emerged four billion years ago, and it is there that life continues to teem today in many strange forms. This blue planet would be better dubbed Oceanus.

The ocean has long been mysterious, its interior largely inaccessible. And although it may not hold the sea monsters that mariners once envisioned, it continues to hold many questions for scientists. Researchers have studied less than 10 percent of the ocean and, because of the difficulty of getting safely to the bottom, have explored no more than 1 percent of the deep ocean floor. Marine biologists know most about the near-shore environments—the coasts, the coral reefs, the kelp forests—and a few other areas that divers can study with ease. But researchers remain ignorant about many aspects of oceanic ecosystems, particularly about life in the midwaters—those between the light-filled upper 100 meters (328 feet) and the near-bottom realm of the deep sea.

From what investigators do know, it is clear that marine animals display a greater diversity of body types than land animals do. Their scientific description requires more broad categories—that is, more phyla (the second most general taxonomic grouping)—than are needed to categorize their terrestrial cousins. Of the 33 animal phyla, 30 describe residents of the ocean, 15 exclusively so. Only 16 phyla include animals found on land or in freshwater—and of those, only one is exclusively terrestrial. This phenomenon reflects the fact that life evolved in the sea and that few life-forms were able to adapt to the absence of water around their bodies.

Yet at the species level, the reverse appears to be true. One and a half million terrestrial species have been described—mostly insects and vascular plants—but total estimates range from five million to more than 50 million. Of the organisms that live in the ocean, however, only 250,000 species have been identified; total estimates run closer to 400,000 to 450,000. This count may change considerably once scientists get a better grasp of life on the ocean floor: some experts posit that between one million and 10 million benthic species have yet to be described.

WATERY PROPERTIES

From people's often terrestrially biased perspective, marine organisms can seem odd. Some of these creatures glow in the dark, many are soft and boneless, and most saltwater plants grow fast and die young—unlike trees, which live to a ripe old age. These differences have arisen because of the physical and chemical characteristics of the ocean.

Seawater is about 800 times as dense as air and is much more viscous. Therefore, marine organisms and particles of food can float endlessly through the water—whereas no creatures drift permanently in the air. Because small life-forms and organic particles are constantly wafting about, some sea animals spend their lives fixed in place, grazing on food in the water around them; on land, only spiders achieve anything like this sedentary lifestyle. The density of water also buoys up organisms, obviating the need for structural supports of cellulose or bone to counteract gravity.

Life underwater has a unique hue as well. Water absorbs light differently than air does. Shorter wavelengths—such as those of the blues and greens—penetrate more deeply than the longer wavelengths of the reds and yellows do. So the view 10 meters below the surface is mostly blue. A few hundred meters deeper there is no sunlight at all and hence no photosynthesis. The midwater and deep-sea communities must depend on the photosynthesizers that reside in the sunlight-filled surface waters. As they sink, these microscopic phytoplankton, zooplankton and decaying particles sustain the fauna of the deep sea.

This rain of plant food is hardly constant, however. Phytoplankton are seasonal and vary by region. Most of the larger species—the ones that turn the ocean green or brown or red when they bloom—thrive in coastal and certain equatorial areas where nutrients are abundant. Much smaller species—called prochlorophytes—are found in tropical and mid-ocean waters. Bottom-dwelling large algae, such as kelp, and seed plants, such as surf grasses, are confined to such a restricted shallow zone around the continents and islands that they contribute little to the overall biological productivity of the ocean, which is relatively modest.

The ocean does not contain much plant life, because concentrations of critical nutrients are lower than they are on land. Phosphorus and nitrogen, for example, are present at only 1/10,000 of their concentration in fertile soil. As a consequence, the ocean supports only a small fraction of what can be grown on reasonably productive land. One cubic meter of soil may yield 50 kilograms (110 pounds) of dry organic matter a year, but the richest cubic meter of seawater will yield a mere five grams of organic matter in that same interval.

The global distribution of what few nutrients there are depends largely on the temperature stratification of the ocean. In the tropics, surface waters are always quite balmy; in temperate regions, these upper waters warm in the summer and are cold the rest of the year. Below the well-mixed surface layer is a narrow zone—called the thermocline—that separates the warm surface from the colder, and thus heavier, water beneath. (An exception to this common configuration occurs near the poles, where the upper and lower levels of the ocean are equally cold.)

It is this cold, heavy water that is the key to the food chain. Because it receives a constant rain of organic detritus from above, deep, chilly water is richly supplied with nutrients. And because no light reaches it, no photosynthesis takes place there—so few organisms take advantage of this abundant nourishment. In contrast, surface water is often barren of nutrients because the sun-loving photosynthesizers have depleted them.

In the tropics, the separation between the warmth at the surface and the cold at depth is so great that even hurricanes and typhoons cannot completely mix the two. As a result, the waters of the tropics remain bereft of nutrients and of the phytoplankton that depend on them. Lacking these clouds of microscopic life, tropical seas normally stay crystal-clear. In temperate regions, winter storms can churn up the ocean, bringing some of the nutrients to the surface. In certain places such as coasts, where steady winds blow the warm surface waters offshore, deep waters rise to take their place. Such areas of nutrient-rich water support some of the world's largest fisheries.

UNDER PRESSURE

Temperature and depth also play an important role because these variables control the availability of oxygen. On land, air provides plants and animals with a fairly constant mixture of this life-giving gas: 210 milliliters per liter. In the sea, oxygen enters only at or near the surface. And because most of the water found in the deep ocean originated at the surface in the coldest parts of the world, it sank carrying large amounts of dissolved oxygen. These water masses may spend centuries in the deep sea before they rise again to the surface. But because life is sparse and moves slowly down there, oxygen is rarely depleted. So, strangely, the ocean is often most oxygen depleted at intermediate depths. For example, in certain areas of the Pacific Ocean an "oxygen minimum zone" occurs between 500 and 1,000 meters below the surface. Only a few organisms are adapted to life in this oxygen-poor environment. Most creatures just travel through it quickly on their way to the surface or back down, where the water is richer in oxygen.

Life in that deep realm is under a great burden. Every 10 meters of seawater adds roughly another atmosphere of pressure: at one-kilometer depth the pressure is 100 atmospheres (100 times what people normally experience). In the profoundest ocean trenches, the pressure reaches more than 1,100 atmospheres. Many invertebrates and some fishes can tolerate the trip from one kilometer deep up to the surface—if they do not have gas-filled sacs that expand as they ascend—and can then survive at one atmosphere for years in refrigerated aquariums.

Despite this opportunity to study them in tanks, marine biologists know relatively little about the organisms that live down in those cold, dark regions. Investigators have learned only that the inhabitants of these realms have unusual adaptations that equip them to live in this environment. For this reason, they are some of the most interesting of all oceanic residents.

THE DEEPEST MYSTERY

Recent studies of the deep sea suggest that although the diversity of species is high, their density is quite low. Food for these organisms arrives in the unending shower of organic particles called marine snow—although sometimes a large carcass, a clump of kelp or a waterlogged tree may settle on the seafloor. Of these sources, though, the marine snow is the most important. As it sinks toward the bottom, microbes, invertebrates and fishes feed on it—and so there is less and less to fall downward. This diminishing supply means there are fewer and fewer consumers at greater depth.

Even more important than the meager, uneven supply of food are the effects of pressure. Deep-sea animals and invertebrates with shells tend to be gelatinous and to have sluggish movements. Their shells are poorly developed because it is difficult to accumulate calcium carbonate under high pressure. If the creatures have skeletons, they are lightweight.

Most deep-sea animals are also small. Many midwater fishes, for instance, are no more than 20 centimeters long. But there are exceptions. Giant squid may reach 20 meters. And the largest comb jellies and siphonophores (relatives of the Portuguese man-of-war) live in the midwater zone, where the absence of strong currents and waves enables these delicate animals to achieve astounding proportions. In fact, the longest animal in the world appears to be a siphonophore of the genus *Praya*, which grows to 40 meters in length and is only as thick as a human thumb. Comb jellies can become the size of basketballs, and the mucus house of the giant tadpole-shaped larvacean *Bathocordaeus charon* may be as large as a Great Dane.

Many of these ghostly creatures glow in their dark abode [see "Light in the Ocean's Midwaters," by Bruce H. Robison, pp. 85 – 93]. Bioluminescence can be found in 90 percent of the midwater species of fish and invertebrates, and many deep-sea fishes have relatively large eyes, so they can see by this faint light. The luminosity serves a variety of purposes: to identify and recognize species, to lure potential prey, to startle a predator and to warn mates of dangers. At depths of a few hundred meters, where dim light still penetrates, the light enables some organisms to blend in with the brighter surface and render their silhouettes invisible from below. Other advantages probably exist that scientists have not yet discovered.

Although they are able to flash light, midwater fishes are often black in color, and many of the crustaceans are red. Because red light cannot penetrate into deep water, this color provides excellent camouflage. Some large jellyfish and comb jellies tend to be purple or red as well.

The top carnivores, which roam near the surface, seldom have such tints. Tuna, billfish, whales, dolphins, seals, sea lions and even seabirds often move through well-lit surroundings as they travel sometimes thousands of kilometers every year. Their movements to feed—whether they are going from deep water to the surface or moving around the globe— result in the longest migrations of animals on the planet.

Some of these creatures have come to represent sea life for most people, so it remains amazing how little biologists know about their habits. How do sperm whales dive a kilometer deep to locate and capture giant squid? Do the yellowfin tuna of the tropical and the semitropical Pacific intermingle? Or are they separate genetic stocks? Part of the reason for this paucity of knowledge is that whales and open-water fishes are extremely difficult to study because they roam the world. Some whales, for example, migrate every year to feed in areas of cold upwelling near the poles and then travel again to reproduce in warmer latitudes. These creatures are the living oil tankers of the sea: using their blubber as fuel, they undergo vast fluctuations in weight, sometimes losing 30 percent of their body mass during migration.

RIGHT UNDER OUR FEET

Not surprisingly, the ocean communities and creatures that researchers know best are those nearest shore: coral reefs, sea-grass beds, kelp forests, coastal mangroves, salt marshes, mudflats and estuaries. These areas are the places people fish, dive, dig for clams, observe shorebirds and, when not paying attention, run boats aground. As a result, these habitats are also the ones people have damaged most severely.

Such environments constitute less than 1 percent of the ocean floor by area, but because they are shallow, well lit and adjacent to landmasses, concentrations of nutrients and biological productivity are relatively high. These coastal areas also link saltwater and freshwater environments. Anadromous fishes, such as salmon, striped bass, shad and sturgeon, reproduce in freshwater rivers and streams, but their offspring may spend years feeding in the ocean before they return to complete the cycle. Catadromous fishes, such as the American and European eels, do the opposite, spending most of their lives in freshwater but going to sea to reproduce.

Perhaps the most familiar near-shore communities of all are those of the intertidal zone, which occupies a meter or two between the high- and low-tide marks. This intertidal stratum is inhabited almost exclusively by marine organisms—although deer, sheep, raccoons, coyotes and bears visit occasionally, as do some insects and a wealth of shorebirds. Organisms living there must be able to endure dryness, bright sunlight and severe shifts in temperature during low tide, as well as the mechanical wear and tear of the waves—which can produce forces equivalent to typhoon winds. It is not surprising, then, to find hard-shelled animals that grip rocks or hide in crevices living there: limpets, periwinkles, barnacles and mussels. Intertidal plants and animals usually occupy distinct horizontal bands that become more densely populated and rich in species at the deeper—and therefore more protected—extreme.

The composition of these littoral communities varies with the shoreline. Sandy shores, for instance, are constantly churned by the waves, so no plants or animals can get a grip for long. Instead most inhabitants are found burrowing underneath the surface. Some tiny animals—called meiofauna—actually live in the interstitial spaces between grains of sand.

Weather patterns and seasonal variations also influence the makeup of the intertidal zone. Temperate areas have the most developed intertidal communities because summer fogs often protect creatures from direct sunlight. In contrast, rocky shores in the tropics are usually quite bare—consisting of a few diatoms, coralline red algae, cyanobacteria, chitons and nerites (both of which are mollusks).

Farther offshore sit the "rain forests" of the marine world: kelp beds and coral reefs. These ecosystems are mutually exclusive but similar in some ways. Both require abundant sunlight and grow within 30 meters or so of the surface. Both contain dominant species that provide a massive, three-dimensional foundation for the community—giant kelp and reef-building corals, respectively. And both house a vast number of species, although coral reefs surpass kelp forests in this regard.

Despite these similarities, their differences are also dramatic. Coral reefs are almost exclusively confined to the tropics, where sea-surface temperatures do not fall below 18 degrees Celsius (about 64 degrees Fahrenheit). Kelp forests do poorly in waters this warm; they are best adapted to temperatures between six and 15 degrees C.

Kelp forests are dominated by the large, brown algae for which they are named. The giant kelp (*Macrocystis pyrifera*) can reach 60 meters in length, stretching 30 meters from the seafloor to the surface and then floating to create a thick canopy. Kelp grow very quickly—as much as half a meter a day in some places. Ninety percent of this plant matter is eaten immediately or washes away to the beach or deep sea, where herbivores later consume it.

These aquatic trees soften the waves and currents and provide food and shelter for many kinds of fish and invertebrates. They are principally grazed by sea urchins and abalone, marine invertebrates that are delicacies for humans and sea otters alike. In some years the urchins get the upper hand, eating the local kelp and other algae—and some invertebrates—to near extinction. It may take several years before the giant kelp can reestablish itself. But in areas where sea otters abound, the urchins are usually kept in check. Indeed, before humans began to hunt for sea otters in the 18th and 19th centuries, populations of urchins and abalones probably never reached the sizes that have supported contemporary commercial fisheries.

Human actions have also profoundly affected many coral ecosystems. These communities are built by stony scleractinian corals, by gorgonians (sea whips and sea fans) and, in the Caribbean, by the hydrozoan fire corals. Scleractinian corals are found in all oceans at a variety of depths. But only the tropical, colonial species construct shallow reefs. These species have photosynthetic dinoflagellates (called zooxanthellae) in their gastric tissues—indeed, 80 percent of corals' soft parts can be made of these creatures. The zooxanthellae photosynthesize and provide the corals with food. These symbiotic dinoflagellates also trigger the corals' rapid calcification, which in turn provides the foundation of the reef structure.

Most reef corals need clear water and a depth of no more than 30 meters so that sunlight can reach their zooxanthellae. The reefs usually do not support many fleshy algae, because grazers—such as sea urchins, parrot fish, surgeonfish and damselfish—constantly nibble at any plant growth. In the early 1980s the importance of these grazers was demonstrated when a pathogen killed 99 percent of the long-spined sea urchins in the Caribbean and algae grew unimpeded, crowding out the corals.

A WORLD IGNORED

Despite their obvious richness, marine ecosystems have been left out of most discussions about saving biodiversity. Part of the reason is that they are out of sight and, hence, out of mind to many scientists and laypersons alike. Nevertheless, it is important to expand their scope as quickly as possible. Current research suggests that at least 70 percent of the world's fisheries are operating at or beyond sustainable levels [see "The World's Imperiled Fish," by Carl Safina, pp. 135–144], and as human populations grow this pressure will only increase.

The intricate connections between the coastal areas, the surface waters, the midwaters and the deep sea are becoming clearer. If society wants the ocean and its myriad creatures to thrive, people must further study these links—and learn to recognize how human actions can alter, perhaps irrevocably, life in the sea.

FURTHER READING

THE OPEN SEA: ITS NATURAL HISTORY. A. Hardy. Houghton-Mifflin, 1956.
DEEP SEA BIOLOGY. J. D. Gage and P. A. Tyler. Cambridge University Press, 1992.
UNDERSTANDING MARINE BIODIVERSITY. National Research Council. National Academy Press, 1995.
MARINE BIOLOGY: AN ECOLOGICAL APPROACH. Fourth edition. James W. Nybakken. Addison Wesley Longman, 1997.

The Ocean's Invisible Forest

PAUL G. FALKOWSKI

ORIGINALLY PUBLISHED IN AUGUST 2002

Every drop of water in the top 100 meters of the ocean contains thousands of free-floating, microscopic flora called phytoplankton. These single-celled organisms—including diatoms and other algae—inhabit three quarters of the earth's surface, and yet they account for less than 1 percent of the 600 billion metric tons of carbon contained within its photosynthetic biomass. But being small doesn't stop this virtually invisible forest from making a bold mark on the planet's most critical natural cycles.

Arguably one of the most consequential activities of marine phytoplankton is their influence on climate. Until recently, however, few researchers appreciated the degree to which these diminutive ocean dwellers can draw the greenhouse gas carbon dioxide (CO_2) out of the atmosphere and store it in the deep sea. New satellite observations and extensive oceanographic research projects are finally revealing how sensitive these organisms are to changes in global temperatures, ocean circulation and nutrient availability.

With this knowledge has come a temptation among certain researchers, entrepreneurs and policymakers to manipulate phytoplankton populations—by adding nutrients to the oceans—in an effort to mitigate global warming. A two-month experiment conducted early this year in the Southern Ocean confirmed that injecting surface waters with trace amounts of iron stimulates phytoplankton growth; however, the efficacy and prudence of widespread, commercial ocean-fertilization schemes are still hotly debated. Exploring how human activities can alter phytoplankton's impact on the planet's carbon cycle is crucial for predicting the long-term ecological side effects of such actions.

SEEING GREEN

Over time spans of decades to centuries, plants play a major role in pulling CO_2 out of the atmosphere. Such has been the case since about three billion years ago, when oxygenic, or oxygen-producing, photosynthesis evolved in cyanobacteria, the world's most abundant type of phytoplankton. Phytoplankton and all land-dwelling plants—which evolved from phytoplankton about 500 million years ago—use the energy in sunlight to split water molecules into atoms of hydrogen and oxygen. The oxygen is liberated as a waste product and makes possible all animal life on earth, including our own. The planet's cycle of carbon (and, to a large extent, its climate) depends on photosynthetic organisms using the hydrogen to help convert the inorganic carbon in CO_2 into organic matter—the sugars, amino acids and other biological molecules that make up their cells.

This conversion of CO_2 into organic matter, also known as primary production, has not always been easy to measure. Until about five years ago, most biologists were greatly underestimating the contribution of phytoplankton relative to that of land-dwelling plants. In the second half of the 20th century, biological oceanographers made thousands of individual measurements of phytoplankton productivity. But these data points were scattered so unevenly around the world that the coverage in any given month or year remained extremely small. Even with the help of mathematical models to fill in the gaps, estimates of total global productivity were unreliable.

That changed in 1997, when NASA launched the Sea Wide Field Sensor (SeaWiFS), the first satellite capable of observing the entire planet's phytoplankton populations every single week. The ability of satellites to see these organisms exploits the fact that oxygenic photosynthesis works only in the presence of chlorophyll *a*. This and other pigments absorb the blue and green wavelengths of sunlight, whereas water molecules scatter them. The more phytoplankton soaking up sunlight in a given area, the darker that part of the ocean looks to an observer in space. A simple satellite measurement of the ratio of blue-green light leaving the oceans is thus a way to quantify chlorophyll—and, by association, phytoplankton abundance.

The satellite images of chlorophyll, coupled with the thousands of productivity measurements, dramatically improved mathematical estimates of overall primary productivity in the oceans. Although various research groups differed in their analytical approaches, by 1998 they had all arrived at the same startling conclusion: every year phytoplankton incorporate approximately 45 billion to 50 billion metric tons of inorganic carbon into their cells—nearly double the amount cited in the most liberal of previous estimates.

That same year my colleagues Christopher B. Field and James T. Randerson of the Carnegie Institution of Washington and Michael J. Behrenfeld of Rutgers University and I decided to put this figure into a worldwide context by constructing the first satellite-based maps that compared primary production in the oceans with that on land. Earlier investigations had suggested that land plants assimilate as much as 100 billion metric tons of inorganic carbon a year. To the surprise of many ecologists, our satellite analysis revealed that they assimilate only about 52 billion metric tons. In other words, phytoplankton draw nearly as much CO_2 out of the atmosphere and oceans through photosynthesis as do trees, grasses and all other land plants combined.

SINKING OUT OF SIGHT

Learning that phytoplankton were twice as productive as previously thought meant that biologists had to reconsider dead phytoplankton's ultimate fate, which strongly modifies the planet's cycle of carbon and CO_2 gas. Because phytoplankton direct virtually all the energy they harvest from the sun toward photosynthesis and reproduction, the entire marine population can replace itself every week. In contrast, land plants must invest copious energy to build wood, leaves and roots and take an average of 20 years to replace themselves. As phytoplankton cells divide—every six days on average—half the daughter cells die or are eaten by zooplankton, miniature animals that in turn provide food for shrimp, fish and larger carnivores.

The knowledge that the rapid life cycle of phytoplankton is the key to their ability to influence climate inspired an ongoing international research program called the Joint Global Ocean Flux Study (JGOFS). Beginning in 1988, JGOFS investigators began quantifying the oceanic carbon cycle, in which the organic matter in the dead phytoplankton cells and animals' fecal material sinks and is consumed by microbes that convert it back into inorganic nutrients, including CO_2. Much of this recycling happens in the sunlit layer of the ocean, where the CO_2 is instantly available to be photosynthesized or absorbed back into the atmosphere. (The entire volume of gases in the atmosphere is exchanged with those dissolved in the upper ocean every six years or so.)

Most influential to climate is the organic matter that sinks into the deep ocean before it decays. When released below about 200 meters, CO_2 stays put for much longer because the colder temperature—and higher density—of this water prevents it from mixing with the warmer waters above. Through this process, known as the biological pump, phytoplankton remove CO_2 from the surface waters and atmosphere and store it in

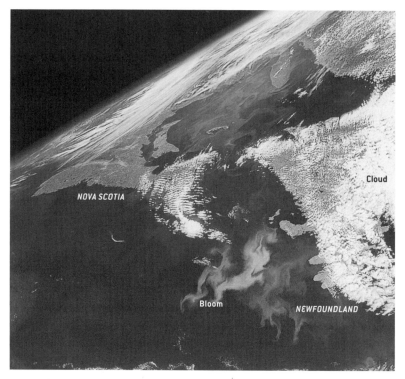

Bloom of phytoplankton called coccolithophores swirls across several hundred square kilometers of the deep-blue Atlantic just south of Newfoundland. Such natural blooms arise in late spring, when currents deliver nutrients from the deep ocean to sunlit surface waters.

the deep ocean. Last year Edward A. Laws of the University of Hawaii, three other JGOFS researchers and I reported that the material pumped into the deep sea amounts to between seven billion and eight billion metric tons, or 15 percent, of the carbon that phytoplankton assimilate every year.

Within a few hundred years almost all the nutrients released in the deep sea find their way via upwelling and other ocean currents back to sunlit surface waters, where they stimulate additional phytoplankton growth. This cycle keeps the biological pump at a natural equilibrium in which the concentration of CO_2 in the atmosphere is about 200 parts per million lower than it would be otherwise—a significant factor considering that today's CO_2 concentration is about 365 parts per million.

Over millions of years, however, the biological pump leaks slowly. About one half of 1 percent of the dead phytoplankton cells and fecal matter settles into seafloor sediments before it can be recycled in the upper ocean. Some of this carbon becomes incorporated into sedimentary rocks such as black shales, the largest reservoir of organic matter on earth. An even smaller fraction forms deposits of petroleum and natural gas. Indeed, these primary fuels of the industrial world are nothing more than the fossilized remains of phytoplankton.

The carbon in shales and other rocks returns to the atmosphere as CO_2 only after the host rocks plunge deep into the earth's interior when tectonic plates collide at subduction zones. There extreme heat and pressure melt the rocks and thus force out some of the CO_2 gas, which is eventually released by way of volcanic eruptions.

By burning fossil fuels, people are bringing buried carbon back into circulation about a million times faster than volcanoes do. Forests and phytoplankton cannot absorb CO_2 fast enough to keep pace with these increases, and atmospheric concentrations of this greenhouse gas have risen rapidly, thereby almost certainly contributing significantly to the global warming trend of the past 50 years.

As policymakers began looking in the early 1990s for ways to make up for this shortfall, they turned to the oceans, which have the potential to hold all the CO_2 emitted by the burning of fossil fuels. Several researchers and private corporations proposed that artificially accelerating the biological pump could take advantage of this extra storage capacity. Hypothetically, this enhancement could be achieved in two ways: add extra nutrients to the upper ocean or ensure that nutrients not fully consumed are used more efficiently. Either way, many speculated, more phytoplankton would grow and more dead cells would be available to carry carbon into the deep ocean.

FIXES AND LIMITS

Until major discoveries over the past 10 years clarified the natural distribution of nutrients in the oceans, scientists knew little about which ocean fertilizers would work for phytoplankton. Of the two primary nutrients that all phytoplankton need—nitrogen and phosphorus—phosphorus was long thought to be the harder to come by. Essential for synthesis of nucleic acids, phosphorus occurs exclusively in phosphate minerals within continental rocks and thus enters the oceans only via freshwater runoff such as rivers. Nitrogen (N_2) is the most abundant gas in the earth's atmosphere and dissolves freely in seawater.

By the early 1980s, however, biological oceanographers had begun to realize that they were overestimating the rate at which nitrogen becomes available for use by living organisms. The vast majority of phytoplankton can use nitrogen to build proteins only after it is fixed—in other words, combined with hydrogen or oxygen atoms to form ammonium (NH_4^+), nitrite (NO_2^-) or nitrate (NO_3^-). The vast majority of nitrogen is fixed by small subsets of bacteria and cyanobacteria that convert N_2 to ammonium, which is released into seawater as the organisms die and decay.

Within the details of that chemical transformation lie the reasons why phytoplankton growth is almost always limited by the availability of nitrogen. To catalyze the reaction, both bacteria and cyanobacteria use nitrogenase, an enzyme that relies on another element, iron, to transfer electrons. In cyanobacteria, the primary energy source for nitrogen fixation is another process that requires a large investment of iron—the production of adenosine triphosphate (ATP). For these reasons, many oceanographers think that iron controls how much nitrogen these special organisms can fix.

In the mid-1980s the late John Martin, a chemist at the Moss Landing Marine Laboratories in California, hypothesized that the availability of iron is low enough in many ocean realms that phytoplankton production is severely restricted. Using extremely sensitive methods to measure the metal, he discovered that its concentration in the equatorial Pacific, the northeastern Pacific and the Southern Ocean is so low that phosphorus and nitrogen in these surface waters are never used up.

Martin's iron hypothesis was initially controversial, in part because previous ocean measurements, which turned out to be contaminated, had suggested that iron was plentiful. But Martin and his coworkers pointed out that practically the only way iron reaches the surface waters of the open ocean is via windblown dust. Consequently, in vast areas of the open ocean, far removed from land, the concentration of this critical element seldom exceeds 0.2 part per billion—a fiftieth to a hundredth the concentrations of phosphate or fixed inorganic nitrogen.

Historical evidence buried in layers of ice from Antarctica also supported Martin's hypothesis. The Vostok ice core, a record of the past 420,000 years of the earth's history, implied that during ice ages the amount of iron was much higher and the average size of the dust particles was significantly larger than during warmer times. These findings suggest that the continents were dry and wind speeds were high during glacial periods, thereby injecting more iron and dust into the atmosphere than during wetter interglacial times.

Martin and other investigators also noted that when dust was high, CO_2 was low, and vice versa. This correlation implied that increased deliv-

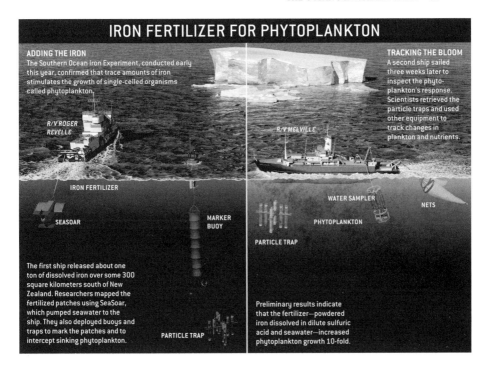

IRON FERTILIZER FOR PHYTOPLANKTON

ADDING THE IRON
The Southern Ocean Iron Experiment, conducted early this year, confirmed that trace amounts of iron stimulates the growth of single-celled organisms called phytoplankton,

TRACKING THE BLOOM
A second ship sailed three weeks later to inspect the phytoplankton's response. Scientists retrieved the particle traps and used other equipment to track changes in plankton and nutrients.

R/V ROGER REVELLE

R/V MELVILLE

IRON FERTILIZER

WATER SAMPLER

NETS

SEASOAR

MARKER BUOY

PHYTOPLANKTON

PARTICLE TRAP

The first ship released about one ton of dissolved iron over some 300 square kilometers south of New Zealand. Researchers mapped the fertilized patches using SeaSoar, which pumped seawater to the ship. They also deployed buoys and traps to mark the patches and to intercept sinking phytoplankton.

PARTICLE TRAP

Preliminary results indicate that the fertilizer—powdered iron dissolved in dilute sulfuric acid and seawater—increased phytoplankton growth 10-fold.

ery of iron to the oceans during peak glacial times stimulated both nitrogen fixation and phytoplankton's use of nutrients. The resulting rise in phytoplankton productivity could have enhanced the biological pump, thereby pulling more CO_2 out of the atmosphere.

The dramatic response of phytoplankton to changing glacial conditions took place over thousands of years, but Martin wanted to know whether smaller changes could make a difference in a matter of days. In 1993 Martin's colleagues conducted the world's first open-ocean manipulation experiment by adding iron directly to the equatorial Pacific. Their research ship carried tanks containing a few hundred kilograms of iron dissolved in dilute sulfuric acid and slowly released the solution as it traversed a 50-square-kilometer patch of ocean like a lawn mower. The outcome of this first experiment was promising but inconclusive, in part because the seafaring scientists were able to schedule only about a week to watch the phytoplankton react. When the same group repeated the experiment for four weeks in 1995, the results were clear: the additional iron dramatically increased phytoplankton photosynthesis, leading to a bloom of organisms that colored the waters green.

Since then, three independent groups, from New Zealand, Germany and the U.S., have demonstrated unequivocally that adding small amounts of iron to the Southern Ocean greatly stimulates phytoplankton productivity. The most extensive fertilization experiment to date took place during January and February of this year. The project, called the Southern Ocean Iron Experiment (SOFeX) and led by the Monterey Bay Aquarium Research Institute and the Moss Landing Marine Laboratories, involved three ships and 76 scientists, including four of my colleagues from Rutgers. Preliminary results indicate that one ton of iron solution released over about 300 square kilometers resulted in a 10-fold increase in primary productivity in eight weeks' time.

These results have convinced most biologists that iron indeed stimulates phytoplankton growth at high latitudes, but it is important to note that no one has yet proved whether this increased productivity enhanced the biological pump or increased CO_2 storage in the deep sea. The most up-to-date mathematical predictions suggest that even if phytoplankton incorporated all the unused nitrogen and phosphorus in the surface waters of the Southern Ocean over the next 100 years, at most 15 percent of the CO_2 released during fossil-fuel combustion could be sequestered.

FERTILIZING THE OCEAN

Despite the myriad uncertainties about purposefully fertilizing the oceans, some groups from both the private and public sectors have taken steps toward doing so on much larger scales. One company has proposed a scheme in which commercial ships that routinely traverse the southern Pacific would release small amounts of a fertilizer mix. Other groups have discussed the possibility of piping nutrients, including iron and ammonia, directly into coastal waters to trigger phytoplankton blooms. Three American entrepreneurs have even convinced the U.S. Patent and Trademark Office to issue seven patents for commercial ocean-fertilization technologies, and yet another is pending.

It is still unclear whether such ocean-fertilization strategies will ever be technically feasible. To be effective, fertilization would have to be conducted year in and year out for decades. Because ocean circulation will eventually expose all deep waters to the atmosphere, all the extra CO_2 stored by the enhanced biological pump would return to the atmosphere within a few hundreds years of the last fertilizer treatment. Moreover, the reach of such efforts is not easily controlled. Farmers cannot keep nutrients contained to a plot of land; fertilizing a patch of turbulent ocean water is even less manageable. For this reason, many ocean experts argue that that once initiated, large-scale fertilization could produce long-term damage that would be difficult, if not impossible, to fix.

Major disruptions to the marine food web are a foremost concern. Computer simulations and studies of natural phytoplankton blooms indicate that enhancing primary productivity could lead to local problems of severe oxygen depletion. The microbes that consume dead phytoplankton cells as they sink toward the seafloor sometimes consume oxygen faster than ocean circulation can replenish it. Creatures that cannot escape to more oxygen-rich waters will suffocate.

Such conditions also encourage the growth of microbes that produce methane and nitrous oxide, two greenhouse gases with even greater heat-trapping capacity than CO_2. According to the National Oceanic and Atmospheric Administration, severe oxygen depletion and other problems triggered by nutrient runoff have already degraded more than half the coastal waters in the U.S., such as the infamous "dead zone" in the northern Gulf of Mexico. Dozens of other regions around the world are battling similar difficulties.

Even if the possible unintended consequences of fertilization were deemed tolerable, any such efforts must also compensate for the way plants and oceans would respond to a warmer world. Comparing satellite observations of phytoplankton abundance from the early 1980s with those from the 1990s suggests that the ocean is getting a little bit greener, but several investigators have noted that higher productivity does not guarantee that more carbon will be stored in the deep ocean. Indeed, the opposite may be true. Computer simulations of the oceans and the atmosphere have shown that additional warming will increase stratification of the ocean as freshwater from melting glaciers and sea ice floats above denser, salty seawater. Such stratification would actually slow the biological pump's ability to transport carbon from the sea surface to the deep ocean.

New satellite sensors are now watching phytoplankton populations on a daily basis, and future small-scale fertilization experiments will be critical to better understanding phytoplankton behavior. The idea of designing large, commercial ocean-fertilization projects to alter climate, however, is still under serious debate among the scientific community and policymakers alike. In the minds of many scientists, the potential temporary human benefit of commercial fertilization projects is not worth the inevitable but unpredictable consequences of altering natural marine ecosystems. In any case, it seems ironic that society would call on modern phytoplankton to help solve a problem created in part by the burning of their fossilized ancestors.

FURTHER READING

AQUATIC PHOTOSYNTHESIS. Paul G. Falkowski and John A. Raven. Blackwell Scientific, 1997.

THE GLOBAL CARBON CYCLE: A TEST OF OUR KNOWLEDGE OF THE EARTH AS A SYSTEM. Paul G. Falkowski et al. In *Science,* Vol. 290, pages 291–294; October 13, 2000.

THE CHANGING OCEAN CARBON CYCLE: A MIDTERM SYNTHESIS OF THE JOINT GLOBAL OCEAN FLUX STUDY. Edited by Roger B. Hanson, Hugh W. Ducklow and John G. Field. Cambridge University Press, 2000.

Ocean primary productivity distribution maps and links to satellite imagery are located at http://marine.rutgers.edu/opp/index.html

Details about the Southern Ocean Iron Experiment are located at www.mbari.org/education/cruises/SOFeX2002/index.htm

Light in the Ocean's Midwaters

BRUCE H. ROBISON

ORIGINALLY PUBLISHED IN JULY 1995

The most expansive animal habitat on the earth lies between the sea surface and the floor of the deep ocean basins. Within this enormous volume live the largest and perhaps most remarkable biological communities anywhere. Yet because this region is so foreign to the world of normal human experience, we still know extraordinarily little about its fauna. But the quest to understand the nature and behavior of these unfamiliar organisms has been making steady progress. Over the past few years my colleagues and I at the Monterey Bay Aquarium Research Institute in northern California have been able to explore the ocean below the sunny surface waters and to examine local ecology from the novel perspective that modern oceanographic technology affords. And, as is often the case when one gets to view something from an entirely new vantage point, that undersea world looks very different from what we had imagined.

My studies of the biology of the ocean's midwaters—a zone that reaches from about 100 meters to a few kilometers below the surface—have benefited enormously from countless hours spent on board *Deep Rover,* a one-person research submarine. Less adventurously but just as effectively, my work has also taken advantage of a remotely operated vehicle (or ROV) named *Ventana,* a maneuverable, computerized platform about the size of a small car that is fitted with an arsenal of cameras, instruments, sensors and samplers.

These two underwater vehicles boast capabilities that far surpass the relatively crude tools that supported previous midwater research. During the 1950s, for example, the marine biologist Eric G. Barham of Stanford University also examined the ocean near Monterey Bay, but at that time he was limited to using sonar and trawl nets towed behind a ship to identify and track the movements of midwater fauna. In the course of his pioneering studies he uncovered a rather limited set of animals—shrimps, lanternfish, squids and arrow worms—and determined the broad patterns of their vertical migrations, from depths of around 300 meters during the day, up to the surface layers at night.

But with the primitive technology then available, Barham's early research missed a tremendous amount of detail in the ocean simply because he could not view it directly. With *Deep Rover* and *Ventana* my colleagues and I have found that the ocean's midwaters contain a far greater variety of organisms than Barham could possibly have caught in his nets: some forms of sea life are simply too fragile to be extracted from their supportive, watery environment. In many respects, we now think of this delicate marine life as *forming* much of that midwater environment.

Among the larger pieces of biological substratum pervading this region are the bodies of gelatinous animals, along with their extended feeding structures and discarded body parts. The most striking contributions of this kind in Monterey Bay are generated by the elongate siphonophores, linear assemblages that can stretch as much as 40 meters—making them some of the longest creatures on the earth. Whether these animals should be regarded as organized colonies of individuals or as a single, complex superorganism remains unclear. I think of them as living drift nets.

Another part of the biological backdrop common in midwater is composed of the balloonlike feeding filters of animals called appendicularians. The most prominent examples are those produced by the giant form, *Bathochordaeus,* an animal that secretes sheets of mucus that look to an underwater observer like floating islands. Because a multitude of midwater animals regularly cast off feeding structures and other body parts, at times the water can become thick with them.

The best way to visualize the midwater environment might be to imagine a dim, weightless world filled with ragged, three-dimensional spiderwebs. Although my colleagues and I have made a host of surprising discoveries about this wispy realm during our explorations, perhaps the most intriguing result to emerge from these efforts to probe the ocean's darkness has been an appreciation for the role of light.

LIFE IN THE TWILIGHT ZONE

Marine biologists had for decades believed that sunlight could penetrate perhaps 300 to 400 meters below the surface of the sea before it became too weak to support vision—a belief they held despite their knowledge that fishes and squids with large, highly developed eyes lived at depths below these levels. But now that we have been able to observe denizens of supposedly dark parts of the ocean, it is becoming clear that these animals are in fact influenced by the tiny amount of sunlight that does filter down to their abode.

Not until I was able peer directly into this world could I begin to appreciate what the midwater habitat is really like. Submerged alone in *Deep Rover* more than half a kilometer below the surface, I have often switched off the lights of the submarine and looked out at the blackness that surrounds the vehicle's transparent passenger sphere. After letting my eyes fully adjust, I can perceive only that looking up is somewhat less dark than looking down. Yet it has become clear to marine biologists that a variety of animals must utilize this subtle difference. Moreover, we have become keenly aware that most creatures of this twilight world are able to augment the scant sunlight reaching them with another form of natural illumination, bioluminescence.

Although bioluminescence is a relatively rare phenomenon in terrestrial ecosystems, the vast majority of the animals that inhabit the upper kilometer of the ocean are capable of producing light in one way or another. Moreover, much of the particulate matter and biological detritus that floats suspended in these waters will glow after it is physically disturbed. These effects can interrupt the normal blackness of the deep ocean with an eerie light.

Midwater animals employ bioluminescence in myriad ways. Some use it as a burglar alarm, coating an advancing predator with sticky, glowing tissue that makes the would-be attacker vulnerable to other visually cued hunters—like bank robbers marked by exploding dye packets hidden in stolen currency. Others use bioluminescence as camouflage. The glow generated by light-producing organs, called photophores, on the undersides of some fishes and squids acts to countershade them: the weak downward lighting effectively erases the shadow cast when the animal is viewed from below against lighted waters above.

The midwater squids *Chiroteuthis* and *Galiteuthis,* for example, clearly demonstrate this use of bioluminescence. Their bodies are transparent except for their dense eyes and ink gland. Ornate light organs arrayed underneath these opaque structures shine downward to countershade them, whatever the position of the squid—head up, head down, inverted or upright. I have found it a bit unnerving to stare eyeball to eyeball with a creature that can pivot its body around a rigid eye that neither blinks nor changes orientation.

Although marine biologists have been able to understand the usefulness of countershading, other examples of bioluminescence have long eluded our logic. One such enigma is a newly discovered species of to-mopterid worm, an active, agile swimmer that has a multitude of paired legs along its tapered body. From specimens caught with nets, biologists have known that some species have structured light organs at the ends of their legs, but only last year James C. Hunt of the University of California at Los Angeles (as well as the Monterey Bay Aquarium Research Institute) and I found a new form of bioluminescent display in a tomopterid that has pigmented pores in roughly the same location as typical leg photo-phores. This species is a "spewer": when stimulated, it squirts a biolu-minescent fluid from each of its leg pores. The discharge forms a lumi-nous cloud that can completely enshroud the body of the worm or leave a glowing trail as it races away. A thimbleful of the ejected fluid contains hundreds of tiny rods that glow brightly yellow. Other types of spewers are known; their strategy may be to cause a visual distraction. But this species remains puzzling. What is the purpose of the display? Why are the tiny light sources rod-shaped? Why is the light given off colored yel-low when most midwater animals have eyes that are sensitive only to blue-green?

Another mysterious application of bioluminescence involves much of the suspended particulate matter and most of the larger gelatinous ani-mals living in midwater: they produce light when stimulated mechani-cally. "Contact flashing" can happen throughout a large volume of this otherwise dim habitat. Most of the time, the surroundings remain tran-quil, with abundant flashers at rest in the dark. But the disturbance of driving Deep Rover through these depths of the ocean can trigger a barrage of exploding lights. The scene underwater can quickly begin to resemble something out of a Star Wars movie.

The natural movements of animals can also cause the ambient biologi-cal lighting to turn itself on, and such bioluminescent responses, when they occur on a large scale, can lead to one of the most remarkable sights in midwaters: a propagated display. This phenomenon starts with local motion triggering contact flashers to fire; these bursts then elicit further flashes like an echo through the adjacent water. Previously poised animals begin moving when the background begins to glow, and their wakes in turn stir up even more light. If contact flashing occurs within a layer of dense particles, the cumulative effect of this bioluminescent activity can look like heat lightning rippling through a cloudy summer night. What-ever the motivation for contact flashing among simpler organisms, more highly developed animals of the midwater region seem well adapted to the situation.

MIDWATER ATTACKERS

Fish such as hake, as well as some squids, are fast-moving, wide-ranging predators, but they often linger near *Ventana,* attracted to the lights of the ROV. It may be that they misinterpret the illuminated waters as an indication that moving prey are present. Perhaps they are conditioned by the daily excursions of sunlight-shunning species that venture near the surface only at night. Such vertical migrations must be light-provoking events, as these animals pass through resident layers of contact flashers. But the potential for movement-induced bioluminescence probably inhibits overall activity, keeping the midwater environment relatively static. Avoiding unnecessary light shows that would give away their position may be the reason mobile animals seem often to remain "parked" in one position much of the time.

Even some predators stay largely motionless. For instance, paralepidids—slender, speedy fish with bodies that look as though they are made of quicksilver—spend the daylight hours standing on their tails, with their sharp snouts thrust upward and their large eyes staring into the waters above. My colleagues and I believe they are searching for silhouettes of their prey against the weakly luminous backdrop. The hatchetfish *Argyropelecus* is another shadow stalker; it has a heavy keel to keep its body horizontal and to stabilize a pair of tubular eyes positioned on top of its body so that its view remains directed upward. *Argyropelecus* lives between about 300 and 600 meters below the surface, where the sunlight must be sufficient to cast perceptible shadows. But a close relative, *Sternoptyx,* lives at depths too great to employ this tactic and has smaller, normally shaped eyes aimed out to the sides.

Further evidence indicates that the weak sunlight of the midwaters is strong enough to guide predators: a diversity of animals living at these depths are transparent. Such a form of appearance (or rather, disappearance) is good protection in this monochromatic, low-light environment. Another optical defense mechanism is red body pigment; this color absorbs the available blue-green light and reflects nothing, a kind of "visual stealth" strategy.

It is not surprising that such optical ploys can work effectively. The visual regime in the midwaters is a bit like the scene from a low-light video surveillance camera. The range of color is narrow; sensitivity is high, but resolution is low; and the directionality of light imparts a flatness to perceptions. To the unaided human eye, the visual field amounts to a coarse pattern of silhouettes and shadows. Within this light-limited milieu there appear to be only four basic shapes: streaks, blobs, strings and spots. Each of these phantoms characterizes a certain kind of subject. Streaks correspond to fishes and squids. Rounded or amorphous blobs are usually gelatinous creatures such as medusae and ctenophores or the weblike feeding structures built from mucus by appendicularians. Stringy material is typically sinking mucus or the tails of siphonophores. Spots can be tiny zooplankton or particles of diffuse organic matter called marine snow.

Within this framework we see a common behavior pattern employed by a variety of creatures. When startled or threatened, some animals change their apparent shape, usually from elongate to rounded. Fishes such as eelpouts curl up into circles and hang motionless in the water. I believe this behavior is a form of mimicry: the animals adjust their appearance to resemble unpalatable objects. From *Deep Rover* I have seen hake strike at fleeing fish while ignoring those that had curled up nearby. The balled-up fish probably resembled medusae—creatures of relatively low nutritional value that deter predators with stinging tentacles. Not all marine biologists agree with this hypothesis, but the observation that this behavior is rarely seen at greater depths (where there is insufficient light for the formation of even rough images) supports the argument for the utility of shape-changing. Such behavior has certainly fooled me at times.

LIGHT FOR THE BLIND

Most gelatinous animals, such as medusae, lack eyes and thus cannot form images of any kind. Yet some of these creatures are clearly sensitive to the lights of *Ventana,* even at a distance, showing a mild dislike for the brightness. My colleagues and I are accumulating evidence that suggests this sensitivity to light may regulate the animal's depth during the day. Changing light levels are known to control the morning and evening migrations of fishes and krill, and it would now seem possible that even eyeless creatures may somehow perceive the sun's presence above them.

We documented one example of such light sensitivity during an encounter with an animal called *Bathyphysa*. This bizarre creature, which is about two meters long, has appeared in front of *Ventana*'s cameras only once, while the vehicle was cruising 500 meters below the surface. When the ROV approached it, the stem of the animal was vertical, with its gas-filled "pneumatophore" uppermost. The stem of the *Bathyphysa* had a mane of elongate, serial stomachs (so-called gastrozooids), each with a probing mouth at its end, and all were writhing like snakes. Several five- to 10-meter-long feeding tentacles radiated out from a round, contracted part of the stem at its center. The stem was exceptionally elastic, a trait that seemed to be explained when we discovered the animal's escape response. Sensing the lights of the ROV, this creature began a series of pounding contractions and relaxations of the upper stem that had the effect of driving the animal downward. In concert with these pulsations, gastrozooids were cast off and left to drift away, one at a time. The result was a determined descent, although a fairly slow and taxing one.

Such episodes suggest that eyeless creatures might well be able to sense even low-level light. In any case, it is clear that they can generate it. *Colobonema,* for example, is a beautifully iridescent little medusa that has a "bell" that is about the size of a coin. In the lights of the ROV, muscle bands in the bell have a blue-green metallic sheen. The medusa's tentacles show a deep blue along their length and brilliant white at the tips.

A fully developed individual has 32 tentacles arrayed uniformly around the base of the bell. Often, however, specimens show fewer appendages set in tiers of different lengths. This appearance is perhaps explained by the animal's behavior: when startled, *Colobonema* darts away, leaving a group of bright, swirling tentacles in its wake. From *Deep Rover* I have observed that the release is occasionally preceded by ripples of luminescence pulsing rapidly through the bell. The many tentacles are then dropped as the bell goes dark and zigzags away into the surrounding blackness.

OCCUPANTS OF THE OXYGEN MINIMUM

One of the characteristic features of the Pacific Ocean near Monterey Bay is a zone that is depleted in dissolved oxygen. Just below the sea surface, oxygen concentrations are close to saturation (that is, the water holds as much oxygen as can possibly be dissolved), but deeper in the ocean, oxygen content diminishes. At about 700 meters of depth, oxygen concentration falls to a value that is only one thirtieth of that near the surface. Below this level is a sharp transition from relatively clear water to a milky layer of very small particles. The milky layer shows a moderate amount of oxygen, and at 1,000 meters the concentration rises further. Within the zone of lowest oxygen near 700 meters resides a unique group of animals that have adapted to meet the physiological challenges of near-anoxia.

One of the most curious inhabitants of the oxygen minimum is the archaic cephalopod *Vampyroteuthis infernalis,* a distant cousin to octopus and squid. A big *Vampyroteuthis* has the size and shape of a soft football. Its body is velvety brown with large eyes that glow like blue opals in the ROV lights. Near the tip of the stubby, conical mantle are two rounded fins and two large light organs with irislike shutters. *Vampyroteuthis* has eight arms like an octopus, but they support a broad web between them. In addition to having suckers, the arms bear a series of paired, fingerlike protrusions, called cirri, that project inward. *Vampyroteuthis* also has two additional appendages: long, elastic sensory filaments that withdraw into pockets between the third and fourth arms on each side.

This creature can be regarded as a living fossil, a modern-day representative of the cephalopods that preceded the evolutionary split into eight- and 10-armed groups. *Vampyroteuthis* propels itself with jets of water expelled from its siphon and by flapping its stubby fins. At the center of the webbed arms is a dark, hooked beak. We do not yet know what this animal eats, but it substantially reduces its own chances of being consumed by living in an inhospitable, anoxic part of the ocean.

My colleagues and I have discovered that this strange animal has a bioluminescent organ at the tip of each of its arms. *Vampyroteuthis* somehow uses these light sources by swinging its webbed arms upward and over the mantle, which turns the suckers and cirri outward and changes the animal's likeness from a football into a spiky pineapple with a glowing top.

This maneuver covers the animal's eyes, but the webbing between tentacles is apparently thin enough for it to see through. We have observed this transformation frequently but remain at a loss to explain exactly what function this unusual behavior might serve.

TECHNOLOGY-DRIVEN EXPLORATION

The present length of *Ventana*'s umbilical tether has permitted us to explore a volume of water one kilometer deep with a visual resolution that extends from about one centimeter to several hundred meters. Although this span covers the ranges of a large portion of the region's midwater species, there are still many measurements we cannot yet make. But this situation is changing. Future technical development by engineers at our institute should allow us to probe even deeper. Soon new optical and acoustic sensing systems will let us examine larger volumes from greater distances and so allow us to assess the distribution of midwater animals even more thoroughly.

We expect eventually to have autonomous probes that will leave time-lapse cameras in place so that we can track slowly moving animals around the clock for days at a time. Fast-swimming robotic vehicles will follow mobile animals, allowing us better to observe their feeding and migration patterns. The possibilities for investigation seem endless. Hence, despite the numerous discoveries already made, we must consider our undersea investigations to have just begun—the ocean's depths are so vast, and there is so much more to explore.

FURTHER READING

DEVELOPMENTS IN DEEP-SEA BIOLOGY. Norman B. Marshall. Blandford Press, 1979.

BIOLUMINESCENCE IN THE MONTEREY SUBMARINE CANYON: IMAGE ANALYSIS OF VIDEO RECORDINGS FROM A MIDWATER SUBMERSIBLE. E. A. Widder, S. A. Bernstein, D. F. Bracher, J. F. Case, K. R. Reisenbichler, J. J. Torres and B. H. Robinson in *Marine Biology,* Vol. 100, No. 4, pages 541–551; 1989.

KIYOHIMEA USAGI, A NEW SPECIES OF LOBATE CTENOPHORE FROM THE MONTEREY SUBMARINE CANYON. G. I. Matsumoto and B. H. Robison in *Bulletin of Marine Science,* Vol. 51, No. 1, pages 19–29; July 1992.

MIDWATER RESEARCH METHODS WITH MBARI'S ROV. Bruce H. Robison in *Marine Technology Society Journal,* Vol. 26, No. 4, pages 32–39; Winter 1992.

NEW TECHNOLOGIES FOR SANCTUARY RESEARCH. Bruce H. Robison in *Oceanus,* Vol. 36, No. 3, pages 75–80; Fall 1993.

Manatees

THOMAS J. O'SHEA

ORIGINALLY PUBLISHED IN JULY 1994

Once upon a time, a young maiden was bathing by the banks of a river. Startled by the sight of approaching men, she jumped in, covering her bottom with a fan. Shyness then doomed her to a life in water: the maiden became a manatee, her fan metamorphosing into its distinctive spatulate tail.

So runs a legend from Mali in West Africa, echoing curiously the origins of manatees as mammals who left the land for a life in sea and river. Whereas myriad tales in native cultures from Africa to the Americas tell of manatees, only recently are the beasts yielding their secrets to scientific inquiry. Among their unique adaptations to life as marine herbivores are an unending supply of teeth—constantly replenishing worn ones—and an anomalously low metabolic rate that allows them to fast for up to seven months. Ponderous and slow, manatees have humans as their only enemy.

Manatees belong to the mammalian order Sirenia—so named because (at least to some eyes) they looked like sirens. The aquatic sirenians probably arose in the Old World; their antecedents were terrestrial mammals that also gave rise to elephants, hyraxes and perhaps aardvarks. We do not know what forces of natural selection drove an ancient mammal to exploit the niche of a large marine herbivore. Yet the beginnings of such an ecological strategy can still be seen today. Domestic sheep on islands off Scotland forage for marine algae in the intertidal zone, even swimming between patches; pigs of the Tokelau Islands in the South Pacific habitually forage along coral reefs, wading with heads submerged.

Fossil evidence suggests that Old World sirenians reached an isolated South America in the Eocene or Oligocene era, more than 35 million years ago. The earliest known true manatee (who lacked, however, the dental technology of modern manatees) lived in the middle Miocene era, 13 to 16 million years ago. The sirenian order currently includes three species of manatees in the family Trichechidae, plus their older relative, the dugong, in the family Dugongidae. A second dugongid, the Steller's sea cow, was hunted to extinction within 25 years of its discovery in 1741.

Up to the late Miocene, dugongids exploiting sea-grass meadows colonized the marine waters of the western Atlantic and Caribbean. Manatees, though feeding on both freshwater and saltwater plants, were restricted to the rivers and estuaries of South America. Dugongs are now found only in the warm, shallow parts of the Indian Ocean and the western Pacific. The animals were evidently displaced about a million years ago by upstart New World manatees, which, as paleontologist Daryl P. Domning of Howard University convincingly argues, outchewed their older relatives.

The chewing prowess of manatees derives from the fact that they never run out of teeth. Manatees possess only premolars and molars (one row on either side of the jaw), but these are continuously replaced by new teeth sprouting at the rear of the row—rather like wisdom teeth—and moving forward. The worn-down front teeth drop out, and the bony tissue separating the tooth sockets continuously breaks down and re-forms to allow the new teeth to move forward at roughly one or two millimeters per month. This process takes place throughout life: even the oldest specimens show new cheek teeth erupting at the rear of each tooth row.

Such an adaptation points to an increased abrasiveness of the manatee diet at some earlier time. Horses, for example, evolved high-crowned cheek teeth in response to the emergence of the true grasses, which have an elevated content of sandpapery silica. Currently true grasses are an important part of the diet of all three species of manatees; in contrast, the staple of the exclusively marine dugong is marine angiosperms (sea grasses). The ancient Caribbean dugongids probably also lived on this diet.

Domning notes that the newly abundant true grasses seem to have invaded South American ecosystems in the Miocene. Continental glaciation in the Pliocene and Pleistocene periods lowered the sea level and increased erosion and runoff of sand and soil. Sand deposition likely increased the amount of abrasive material ingested with the sea grasses. The more sand-tolerant trichechids invaded these habitats as well. By about a million years ago they had broadened their feeding niche to include sea grasses and replaced dugongids in the Atlantic and Caribbean.

In the Amazon region, mountain building in the late Miocene era created a transient closed basin; trichechids isolated here became the Amazon manatee (*Trichechus inunguis*). The West Indian manatee (*T. manatus*) is apparently a little-changed descendant of coastal South American trichechids of the Pliocene-Pleistocene time; at present it can be distinguished into subspecies from Florida and the Antilles. A similar form also reached West Africa by way of transoceanic currents comparatively recently—perhaps since the late Pliocene—to give rise to the West African manatee, *T. senegalensis.*

Manatees have had a long and intimate relationship with humans, mostly as food. These large animals—my colleagues and I once weighed a Florida manatee at 3,650 pounds—have been hunted and relished by members of coastal and riverine cultures throughout the tropical Atlantic. In West Africa, manatees are lured into box traps with cassava, speared from platforms on stilts, entangled in nets, shot by harpoons on baited triggers and trapped by fences on outgoing tides. South American Indians place logs across streams to trap manatees in receding waters.

The creatures also feature in numerous superstitions and legends in native cultures. In Mali, manatees in the river Niger are considered evil spirits; only a few tribesmen know the proper incantations to hunt and kill a manatee without dying or going mad. When a hunt is successful, cuts of meat are distributed according to social status. Some other protocols must be followed, too. If a pregnant woman eats certain cuts, the unborn child is believed to be in danger of becoming an adult of low moral character. The oils and skin of the manatee are made into remedies for various ailments, and potions are made from its ribs.

Across the Atlantic, in the headwaters of the Amazon in Ecuador, a Siona Indian shaman told my fellow scientists and me a legend about the origin of the Amazon manatee. An ancient god was deceived and trapped by a tapir, who cruelly subjected the god to attack by piranhas. The god escaped and in revenge banished one of the tapir's daughters to live forever in the water—as a manatee.

Inhabitants of the coasts and rivers of Central America, the Caribbean and northeastern South America prize manatees as food and as a source of medicine. Great stealth and machismo are needed to harpoon manatees from dugout canoes on the Orinoco River. Accordingly, the tiny middle-ear bones of manatees are worn as magical charms against evil and disease. The Piraoa Indians of Venezuela, however, have a taboo on hunting manatees and river dolphins. The Piraoa think they will die if they eat manatee meat; they believe the creatures to be bewitched humans who dwell in underwater cities at the bottom of the Orinoco.

In 1493 Christopher Columbus and his crew became the first Europeans to see manatees in the New World. They reported the creatures as mermaids. In the decades that followed, Europeans acquired a closer ac-

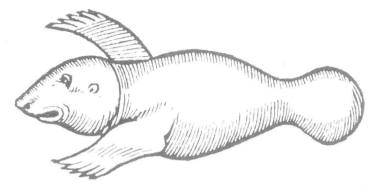

Earliest known representation of a manatee, from Fernández de Oviedo's 1535 *History of the Indies,* is based on Christopher Columbus's description of the creatures. His crew, the first Europeans to see manatees, took them to be mermaids.

quaintance with manatees. Indians used manatee hides as shields against Spanish explorers armed with crossbows. In the late 1600s William Dampier, the English buccaneer, fed his crews with boatloads of manatee meat from Panama, supplied by Miskito Indians.

Later, manatees were exploited commercially, both as bush meat for laborers at South American frontier posts and in processed form for resale in distant markets. Shiploads were exported to the West Indies from Guyana, Suriname and Brazil in the 17th and 18th centuries. Exports ended with the close of the 19th century, when manatee populations had become greatly reduced. Legal trade in Amazon manatees continued in Brazil until 1973. Perhaps as many as 7,000 were killed in peak years of the late 1950s.

The end of commercial exploitation coincided with the rise of other threats to the survival of manatees. By the 1970s it became clear that increasing numbers were dying in accidents and in encounters with man-made artifacts. Inexpensive synthetic gill nets that became widespread in tropical fisheries incidentally entangled and suffocated manatees. Rivers and estuaries became polluted and turbid because of deforestation and erosion, blocking the light needed by aquatic plants. Manatee habitats became endangered, especially in areas where humans are plentiful. An estimated 800 to 1,000 people move to Florida every day; many newcomers settle in coastal areas, where wetlands are drained and replaced by housing—complete with canals and boat docks. Water quality has dropped; in Tampa Bay, 80 percent of the sea-grass beds have vanished over recent decades. In Florida, boat propellers inflicted gruesome wounds.

Deaths of manatees accidentally caused by humans doubled during the 1980s, as the number of registered boats plying the waters also grew dramatically.

Clearly, the manatee needed protection, or else it would rapidly become legend alone. So little was known about its habits and habitat, however, that it was difficult to formulate plans to save the manatee. Then, in the 1970s, the Marine Mammal Protection Act and the Endangered Species Act initiated a spurt of research, concentrated in Florida.

Biologists began, for one thing, to dissect systematically carcasses cast on the beach. The dissections augmented earlier knowledge about the anatomy of manatees. In addition to replacing their teeth constantly, manatees have evolved other features helpful to their aquatic lifestyle.

A manatee's lungs have single lobes and lie above the abdominal cavity—along the back—enabling the creature to remain horizontal under water. Its gastrointestinal tract is long, and the animal digests food in the hindgut. There is also a "cardiac gland" in a pouch off the stomach. The gland contains specialized secretory cells, and the pouch protects them from the abrasive diet.

The animal's heavy, dense bones lack marrow; they help to keep the manatee submerged. Only the backbone contains some marrow for producing red blood cells. The animal can adjust its depth by changing the volume of its lungs. Its lips are large, studded with bristles, and prehensile—forming a kind of abbreviated trunk. The long, paddlelike flippers are used to manipulate food and to walk stealthily along the bottom. Counts of growth layers in ear bones suggest that the manatee's normal life-span is about 60 years.

By examining the stomach contents of manatees, Catherine A. Beck and her colleagues at the National Biological Survey in Gainesville, Fla., have found that the diet has much variety. For instance, manatees eat most local aquatic plants, as well as acorns from overhanging oak trees. Unfortunately, their stomachs also include refuse, such as plastics, condoms, fishing line and steel hooks—some of which have led to their death.

Also in the 1970s biologists began to study live, captive manatees, thereby uncovering some intriguing facts about the rate at which the animals consume energy. The metabolic rate is related to the amount of oxygen a mammal consumes per unit time divided by its body weight. Most species consume energy at rates that depend on their size. For example, small mammals have a large surface area for their weight. Because heat is lost through the surface and mammals have to expend energy to main-

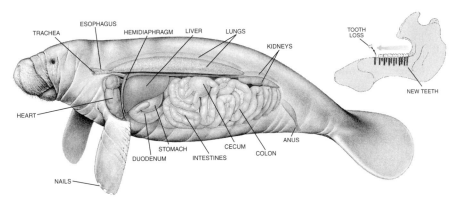

Anatomical drawing of a manatee reveals several adaptations to life as an aquatic grazer. Lungs lying along the back help to keep the animal horizontal under water. Flippers and a prehensile upper lip are used to manipulate food. New teeth (*right*) are constantly generated at the rear of each tooth row and move forward. Eventually, teeth worn out by chewing abrasive grasses drop out. The food is digested as it passes through the long hindgut (cecum to anus) for several days.

tain a constant body temperature, smaller mammals use more energy relative to their size and have higher metabolic rates.

On a mouse-to-elephant curve of metabolic rates for mammals of different weights, most kinds of marine mammals fall where they should— going by their respective sizes. But Blair Irvine of the U.S. Fish and Wildlife Service and C. James Gallivan and the late Robin C. Best of the National Institute for Amazon Research in Brazil found that manatees deviate sharply from this pattern. Amazon manatees metabolize energy at only 36 percent of the rate expected for mammals of their size; in Florida, manatees come in at a mere 15 to 22 percent of the predicted values.

What are the ramifications of such low metabolic rates? One of the most astounding is the capacity that the Amazon manatee has for prolonged fasting. As in earlier days of sirenian evolution, seasonal flooding in the upper Amazon gives rise to large floating meadows of grasses and other vegetation. Amazon manatees swim through the treetops of the flooded forests in times of plenty, but during the dry season they can become isolated for months in lakes and pools devoid of vegetation [see "Flooded Forests of the Amazon," by Michael Goulding; *Scientific American*, March 1993]. Like mammals of colder zones preparing for winter dormancy, the creatures put on large quantities of fat during the flood season, which allows them to survive the lean times of the dry season. Best calculated that Amazon manatees can go without feeding for almost seven months by subsisting on stored fat and by expending energy at their customarily low metabolic rates.

We do not know if there is a similar link between metabolism, fattening and seasonal fasting for other species of manatees. But the influence of low metabolic rates on the distribution of Florida manatees is well known. The animals do not live north of Florida or Georgia on a year-round basis. Unable to metabolize energy fast enough to counter heat lost to cool surrounding waters, manatees probably cannot expand beyond subtropical climes. Nearly all sirenians throughout geologic time have lived in warm regions.

Almost every year, however, manatees are sighted north of their typical range. These summer wanderers return south in autumn or die and wash ashore in Virginia or the Carolinas later in winter. The carcasses are marked by an absence of fat deposits and other signs of long-term exposure to cold. Wanderers have been verified as far north as the lower Potomac; I strongly suspect that the "Chessie" sea monsters of Chesapeake Bay in the 1970s and 1980s were errant Florida manatees.

Even in Florida, manatees find winter temperatures stressful. They respond by migrating to the southern third of the peninsula or to local sources of warmer water. These include artesian springs, such as Crystal River on the Gulf Coast, and effluents of pulp mills and electric power plants. On the coldest days, 300 or so animals aggregate at several of these sites. For many years, winter aggregations gave biologists the only window into the lives of manatees in the wild.

In the early 1950s Joseph C. Moore of Everglades National Park noted that individual manatees can be distinguished by boat-propeller scars; he was thus able to make some basic behavioral observations. His initial study was followed in the 1960s by Daniel Hartman, then a graduate student at Cornell University. Hartman tracked individual females and their offspring at Crystal River. Currently the National Biological Survey maintains a computer-based scar-pattern catalogue identifying hundreds of Florida manatees from sites throughout their range.

Dozens of wild females have now been observed for more than a decade—some for 20 to 25 years—giving us insight into their reproductive cycle. A female attains sexual maturity as young as three years of age and continues to reproduce for more than 20 years. One calf is born every two to three years; there are occasionally even twins.

Social interactions seem to revolve around reproduction. Manatees are basically solitary animals, but when a female is in estrus, she is pursued by a herd of six to 20 jostling and wrestling males. About a year after mating, apparently with several males, the female selects a secluded area for birthing. Mother and calf stay together as a nursing pair for at least a year, maintaining contact by faint, squeaky vocalizations.

A few females and their calves have been seen together for up to a year after weaning and perhaps continue to recognize one another for much longer. Sometimes nursing females will "adopt" and suckle calves that are not their own. Manatees will occasionally socialize in transient groups, which individuals join and leave seemingly at random.

In 1978 speed zones for boats were established at winter aggregation sites. Still, manatees and boats continued to collide at other times and places. If they were to protect the creatures in their diverse habitats, policymakers needed to know more about the distribution and migration patterns of manatees. In the 1980s advances in radiotelemetry provided a means of observing manatees in their travels over long distances.

Manatees in the waters of Brazil and Florida were first to be tracked by radio in the 1970s. Investigators attached a transmitter to a belt—designed to corrode and fall off after the study—around the constriction between a manatee's body and its spatulate tail. The technique was not, however, useful in coastal habitats, where saltwater impedes the passage of radio waves.

So my colleagues Galen B. Rathbun, James P. Reid and James A. Powell designed a floating transmitter attached to the belt by a two-meter-long flexible nylon tether. The tether was equipped with swivels to minimize drag and with breakaway links to prevent the manatee from getting trapped should the tether snag. The device put the antenna in the air during most manatee activities in shallow waters. We could locate animals as far as 20 to 30 miles away from light aircraft and five to 10 miles away from boats or shore.

Soon we were able to take the floating-transmitter concept a step further. Bruce Mate of Oregon State University had been trying to track great whales in the open ocean. He solved this tremendous logistical problem by attaching to the whales "platform-transmitter terminals" monitored by satellites. These transmitters, used to track weather balloons and vessels at sea, emit ultrahigh-frequency signals. Satellites receive the signals and pass on the encoded information to processing centers on the earth. From the centers it travels via telephone links to personal computers. Within hours or less of the last pass of a satellite over a transmitter, a scientist can know a whale's location.

In 1985 Rathbun, Mate and Reid released a manatee with a floating satellite transmitter off Florida's Gulf Coast. Its signals were received by satellites of the National Oceanic and Atmospheric Administration. Since

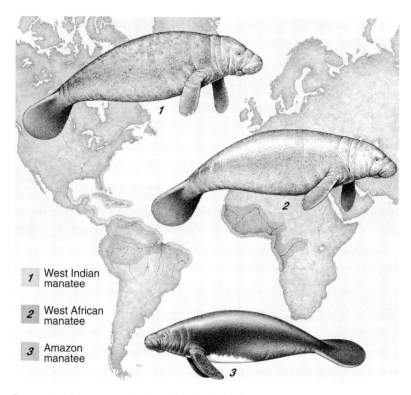

1 West Indian manatee

2 West African manatee

3 Amazon manatee

Three species of manatees now live in tropical and subtropical waters. The West Indian manatee, shown with algae on its back, most closely resembles the earliest trichechid. The African manatee is descended from dispersed West Indian manatees. The Amazon species feeds on grasses and has the most complex dentition.

then, we have tracked more than 100 manatees with tethered transmitters, most by satellite. The technique has been adopted by the Florida Department of Environmental Protection, is being used in Puerto Rico and has been applied to dugongs in Australia.

The satellite link reveals a manatee's location (within 100 meters), the water's temperature and the number of times the transmitter is tipped—giving clues about the animal's activity. In Florida, this information is correlated with maps of sea-grass beds, warm waters and other manatee resources at the Marine Research Institute of the Florida Department of Environmental Protection in St. Petersburg. Earlier, it was believed Florida manatees moved slowly and were essentially nomadic. Now we know they can travel fairly fast—sometimes 50 kilometers a day—and their seasonal movements can be quite directed. Some females, for instance, will often graze in roughly the same areas every summer and head for the same hot spots every winter. Offspring appear to learn the mother's migration patterns.

Generally, though, manatees are flexible in the kinds of habitats they occupy. We have tracked individuals from southern Georgia and northeastern Florida—where the primary forage is salt-marsh grass available only on banks at high tide—traveling in less than five days to the Merritt Island National Wildlife Refuge, where they feed almost exclusively on submerged, rooted marine angiosperms. After lingering in this region, rich in classic Florida wildlife such as wading birds, sea turtles, bottle-nosed dolphins and alligators—which sometimes seize and detach the trailing transmitters—the same manatees may continue southward, arriving in the urban environs of Fort Lauderdale and Miami for the coldest weeks of the year.

Increased knowledge about the adaptability of manatees to diverse habitats and about their rate of reproduction has made us guardedly optimistic about the survival of manatees in Florida—once they are adequately protected. Administrative efforts by the state of Florida, the U.S. Fish and Wildlife Service and local governments to reduce accidents have a high potential for paying off. We have combined the resighting histories of manatees in the scar-pattern catalogue with recent statistical theories to estimate the year-to-year survival rate of adult manatees. Chances of survival are good in areas with solid histories of protection, like Crystal River. The population there has grown from about 60 animals some 20 years ago to nearly 300 now. In Florida as a whole, 1,856 have been counted by air in winter.

We do not know how many manatees were missed in these counts. Although the general pattern from the 1970s through the 1980s pointed to an increase in manatees in several regions of Florida, trends in most recent years leave doubt about whether the population is growing at all. Along with lower estimates of survival in less protected regions (from our mathematical models) and more manatees being found dead from human activities, uncertainties in the recent population data call for continued efforts for conservation focused on some key areas of Florida. Should such efforts maintain their momentum—and barring unforeseen catastrophes—the Florida manatee could become a rare success story for endangered species.

Ultimately what will save these creatures is a sympathetic public. In this regard, there has been some genuine progress. Manatees have become

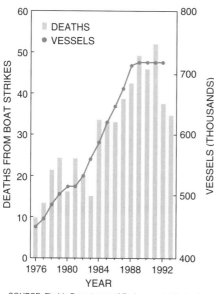

The graph details the increase in the number of boats registered in Florida and in the number of manatees killed by boats since the mid-1970s. The correlation, while not proving a cause-and-effect relationship, is highly suggestive. The rise in the number of boats also indicates that their use for recreation is increasing. Consequently, so is their destructive effect on the manatee habitat.

SOURCE: Florida Department of Environmental Protection

extremely popular. For example, walking into my daughter's elementary school classroom, I was pleasantly surprised to see—in the permanent alphabet displayed above the blackboard—along with "A" for apple, "M" for manatee.

Internationally, the situation is less encouraging. Although most of the 40 to 45 nations in which manatees live offer them legal protection, the laws are not well enforced. Also, few attempts have been made (outside of Florida) to protect their habitat. Guatemala created the world's first manatee reserve some 30 years ago. The secretive animals are rarely glimpsed there, but the reserve is still maintained. Along the coast of Panama, where Dampier provisioned his crews centuries ago, meager numbers of manatees persist in rivers of the Bocas del Toro region, and Panamanian conservation groups are working on their behalf.

It will be an uphill battle. But similar attempts are starting elsewhere. A new generation of conservation biologists from the tropical Atlantic is increasingly enthusiastic and concerned about manatees. The energy of these scientists was evident at the First International Conference on Manatees and Dugongs held at the University of Florida this past spring. Efforts to learn more about manatees in the tropics and to apply this information to conservation seem to be growing, providing seeds of hope for the future.

The Warauno Indians of the Orinoco Delta Territory of Venezuela refer to the Milky Way Galaxy as "the road of the manatee." I remain hopeful that the underwater roads of the manatee will continue to be traveled along our tropical Atlantic rivers and coastlines here on earth, to the marvel and delight of future generations.

FURTHER READING

ECOLOGY AND BEHAVIOR OF THE MANATEE (*TRICHECHUS MANATUS*) IN FLORIDA. Daniel S. Hartman. American Society of Mammalogists, Special Publication No. 5, June 27, 1979.

DISTRIBUTION, STATUS, AND BIOGEOGRAPHY OF THE WEST INDIAN MANATEE. L. W. Lefebvre, T. J. O'Shea, G. B. Rathbun and R. C. Best in *The Biogeography of the West Indies: Past, Present, and Future.* Edited by C. A. Woods. Sandhill Crane Press, 1989.

MANATEES AND DUGONGS. J. E. Reynolds III and D. K. Odell. Facts on File, 1991.

WATERBORNE RECREATION AND THE FLORIDA MANATEE. T. J. O'Shea in *Wildlife and Recreationists: Coexistence through Management and Research.* Edited by Richard L. Knight and Kevin J. Gutzwiller. Island Press, Covelo, Calif.

Secrets of the Slime Hag

FREDERIC H. MARTINI

ORIGINALLY PUBLISHED IN OCTOBER 1998

hump. After an hour of descending through near-total darkness in the research submarine *Alvin,* we slide into the silty ocean bottom roughly 1,700 meters (just over one mile) below the surface of the Pacific, off the coast of southern California. The pilot switches on the floodlights, illuminating a dense cloud of sediment kicked up by our arrival. Several minutes pass as we wait for the debris to settle and activate the sub's sonar system, which shows a large target roughly 240 meters away. As we move closer, we see through *Alvin's* portholes the ghostly white carcass of a 32-metric-ton gray whale. The whale's watery grave is anything but peaceful: it is swarming with hundreds of half-meter-long hagfishes, which are methodically gnawing away at the whale's chalky blubber, bite by bite.

Scenes like this are eerie enough to keep some people up at night—and they change forever one's concept of burial at sea. But for my colleagues and me, who study the biology of hagfishes, they provide a fascinating glimpse into the lives of these strange and slimy animals. For years, the habits of hagfishes—which are sometimes called slime hags—and their place on the evolutionary tree of life have been a matter of conjecture. But recent studies indicate that hagfishes, which appear to have changed little over the past 330 million years, in many ways resemble the first craniates (animals with a braincase). For instance, the evolutionary path leading toward humans—and all other vertebrates (animals with a backbone)—probably diverged from that of hagfishes 530 million years ago. New research also shows that hagfishes are much more abundant—and probably play a much more important role in the ecology of the ocean-bottom community—than anyone would have guessed a decade ago.

SLIME BALLS OF THE SEA

The word "slimy" can only begin to describe the average hagfish: one good-size adult can secrete enough slime from its roughly 200 slime glands to turn a seven-liter bucket of water into a gelatinous mess within minutes. Hagfishes release slime in varying amounts, depending on the circumstances. They tend to produce slime in small amounts while feeding on a carcass, a behavior that might be designed to ward off other scavengers. But when attacked or seized, a hagfish can ooze gobs of goo, from all its slime glands at once. Although the slime is initially secreted as a small quantity of viscous, white fluid, it expands several hundred times as it absorbs seawater to form a slime ball that can coat the gills of predatory fish and either suffocate them or distress them enough to make them swim away. But for all its utility, the slime appears to be equally distressing to the hagfish. To rid its body of the sticky mucus, a hagfish literally ties its tail in a knot and sweeps the knot toward its head to scrape itself clean.

People often mistake the hagfish for an eel because both animals are long and cylindrical. The common names of several species of hagfish even include the term "eel," usually accompanied by a descriptive adjective ("slime eel," for example). As is so often the case, however, such common names are misleading. Hagfishes are not eels at all: true eels are bony fish with the requisite prominent eyes, paired pectoral and pelvic fins, a hard skeleton, bony scales and strong jaws. Like other bony fish, eels rely for respiration on gills that are attached to bones called gill arches and covered by a bony flap called an operculum.

In contrast, hagfishes are much simpler in form and function [*see illustrations on pages 108 and 110–111*]. They lack true eyes and paired fins, and their rudimentary skeleton consists only of a longitudinal stiffening rod made of cartilage, called the notochord, and several smaller cartilaginous elements, including a rudimentary braincase, or cranium. Hagfishes do not have scales; instead they have a thick, slippery skin and large, complex slime glands. In addition, they lack jaws, and their gills are a series of pouches that are different from the gills of any other living fish.

Hagfishes can be found in marine waters throughout the world, with the apparent exception of the Arctic and Antarctic seas. Although the animals always live near the ocean bottom, they can survive at a variety of depths. Water temperature is the primary factor that limits the habitat of hagfishes: they appear to prefer waters cooler than 22 degrees Celsius (71 degrees Fahrenheit). In the cold coastal waters off South Africa, Chile and New Zealand, the animals sometimes enter the intertidal zone, where they have been collected in tide pools as shallow as five meters. In tropical seas, though, hagfishes are seldom seen at depths shallower than 600 meters.

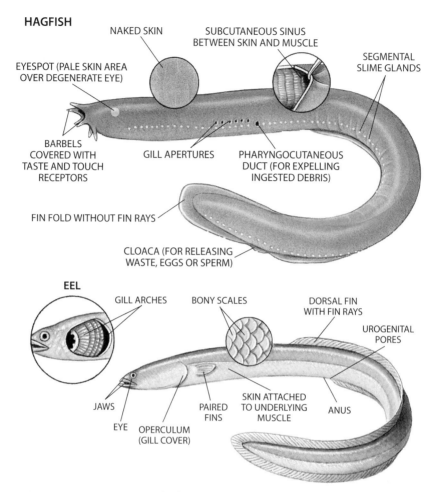

HAGFISH

NAKED SKIN

SUBCUTANEOUS SINUS
BETWEEN SKIN AND MUSCLE

SEGMENTAL
SLIME GLANDS

EYESPOT (PALE SKIN AREA
OVER DEGENERATE EYE)

BARBELS
COVERED WITH
TASTE AND TOUCH
RECEPTORS

GILL APERTURES

PHARYNGOCUTANEOUS
DUCT (FOR EXPELLING
INGESTED DEBRIS)

FIN FOLD WITHOUT FIN RAYS

CLOACA (FOR RELEASING
WASTE, EGGS OR SPERM)

EEL

GILL ARCHES

BONY SCALES

DORSAL FIN
WITH FIN RAYS

UROGENITAL
PORES

JAWS

EYE

OPERCULUM
(GILL COVER)

PAIRED
FINS

SKIN ATTACHED
TO UNDERLYING
MUSCLE

ANUS

External views of a Pacific hagfish (*top*) and an eel of the genus *Anguilla* (*bottom*) reveal that despite their similar shapes, the eel is a more highly evolved, bony fish. Unlike the eel, the hagfish lacks jaws, paired fins, eyes, scales, dorsal fin rays, gill arches and gill covers.

There are roughly 60 species of hagfishes, most of which are members of two major genera: *Eptatretus* or *Myxine*. (Many species in these genera, however, are known only from single specimens.) The genus *Eptatretus*, with roughly 37 species, includes the largest hagfish known, *E. carlhubbsi*, which can reach a length of 1.4 meters and can weigh several kilograms. Underneath their skin, *Eptatretus* species have the evolutionary remnants of eyes that are covered by translucent eyespots. Their heads also bear traces of lateral lines, sensory structures that extend down the sides of bony fish. Individual *Eptatretus* live in long-term burrows in the ocean floor but may roam widely among rocks or other hard substrates.

Members of the genus *Myxine,* which includes roughly 18 species, are more specialized than *Eptatretus* for living in burrows. *Myxine* are generally more slender, have even more degenerate eyes that lack eyespots, and show no traces of lateral lines. Typical *Myxine* live in transitory burrows and are always found in or near soft, muddy sediments.

The feeding habits of hagfishes—which can eat small, live prey and act as scavengers—are particularly distinctive. As a hagfish feeds, it protrudes a very effective feeding apparatus consisting of two dental plates, each supporting two curved rows of sharp, horny cusps. The dental plates are hinged along the midline, allowing them to open and close like a book. To take a bite, a hagfish extends its feeding apparatus, causing the "book" to open, and presses the dental plates against a fleshy surface—whether it be the body of a sea worm, a dead fish or your hand. When the hagfish withdraws the apparatus, the book closes and the opposing cusps grasp and tear the flesh, carrying it into the mouth. (The fang situated above the dental plates keeps live prey from wriggling away between bites.)

This feeding method works quite well when a hagfish preys on thin-skinned, soft-bodied sea worms, but the cusps cannot pierce the scales of fish or the skin of whales. Unless other scavengers have already opened the way, when feeding on a large carcass a hagfish usually takes the easy route, entering the body through the mouth, gills or anus. It then consumes the soft tissues from within, until only the bones and skin remain. More than one disappointed fisher has hauled in a prize fish that turned out to be a hollow shell full of hagfishes.

Only a few details are known about hagfish reproduction. Hagfish gonads form in a fold of tissue on the right side of the abdominal cavity. In a female an ovary forms in the anterior two thirds of the fold; in a male, a testis forms in the posterior one third. Curiously, individuals with both types of gonads are found occasionally. Females, which in some species outnumber males more than 100 to one, produce between 20 and 30 yolky, shelled eggs at a time. There are no oviducts; mature eggs are released into the abdominal cavity. The eggs—which vary in size from 20 to 70 millimeters, depending on the species—usually have hooked filaments at either end that enable them to lock together and be ejected in a chain. In males the testis produces sperm in follicles that release sperm into the abdominal cavity. Eggs or sperm then leave the abdominal cavity through a large pore into the cloaca, an excretory chamber that also receives and expels urinary and digestive wastes.

Apart from these anatomical details, however, the sex lives of hagfishes remain almost a complete mystery. We assume that hagfish females lay their eggs for subsequent fertilization by males, but we have no idea where, when or how this occurs. We also do not know why the sex ratio of the hagfish is biased toward females or how often the females produce eggs.

The embryonic development of the hagfish is also still a black box. Despite more than 100 years of searching, only three fertilized eggs of the genus *Myxine* have ever been found, and those were damaged. The situation is only slightly better for other genera of hagfish: roughly 200 fertilized eggs of the genus *Eptatretus* were collected in California's Monterey Bay between 1896 and 1942, but none have been recovered since.

Many other aspects of hagfish life are equally mysterious. For instance, juvenile *Myxine*—those under 170 millimeters in length—have never been collected. Where are they, and what do they eat? How fast do they grow? At what age do they mature sexually? As yet, we have no answers.

LIVING FOSSILS

Given the bizarre and mysterious biology of hagfishes, it is no wonder that these blind, jawless, scaleless, finless, bottom-dwelling creatures were not immediately recognized (or acknowledged) as distant cousins of humans. In 1758, for instance, noted biologist Carl von Linné, writing under the name of Linnaeus, classified hagfishes as Vermes, or worms (rather than

Anatomical views of the anterior (*left*) and posterior (*right*) parts of a Pacific hagfish highlight both the animal's unique specializations and other, more general characteristics—such as a cranium—that persist in more evolved animals. (The middle—roughly one third of the animal's length—has been omitted.) Like a small proportion of most species of hagfishes, this specimen has both an ovary and a testis.

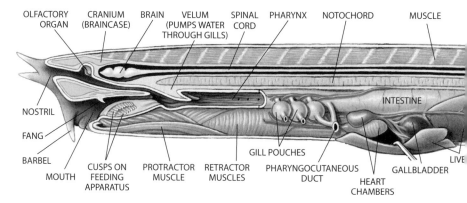

fishes), although we now know that hagfishes and worms are only distantly related.

Today, however, scientists recognize that hagfishes are virtual biological time machines. The term "living fossil" is often used to refer to the coelacanth, a rare, deepwater fish with fleshy, lobed fins that was first caught in 1938 in waters off the east coast of South Africa, among the Comoro Islands. But hagfishes make coelacanths look like evolutionary newborns: coelacanths may have changed little since they first appeared in the fossil record 60 million years ago, but a fossilized hagfish, *Myxinikela,* was found in sediments deposited roughly 330 million years ago. Aside from *Myxinikela*'s large eyes, if it were alive today it could easily pass for a modern hagfish.

Biologists neglected hagfishes for much of the past century primarily because of the way they classified animals. Until relatively recently, they relied on common features, such as the presence or absence of eyes or jaws, to establish relatedness between creatures. Under this scheme, hagfishes were lumped together with lampreys in a group called either Agnatha (literally, "no jaws") or Cyclostomata ("round mouths"). Hagfishes and lampreys were classified together because both lack jaws, paired fins, a bony skeleton and scales. Because the habitats of hagfishes make them relatively hard to come by, biologists concentrated on lampreys, which spend part of their lives in freshwater streams and rivers and therefore are much easier to catch.

In more recent years, the acceptance of phylogenetic systematics, or cladistics—classifying animals according to shared, specialized characteristics—has forced a reevaluation of the old methods for deciding what is related to what. Biologists now recognize that it is impossible to tell

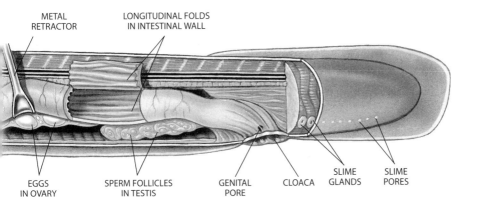

METAL RETRACTOR   LONGITUDINAL FOLDS IN INTESTINAL WALL

EGGS IN OVARY   SPERM FOLLICLES IN TESTIS   GENITAL PORE   CLOACA   SLIME GLANDS   SLIME PORES

whether the ancestors of a given organism never had a particular feature or whether they had the feature but their descendants simply lost it sometime during evolution. Neither hagfishes nor snakes have legs, for example, but that does not mean that they are related. Hagfishes have never had paired fins—let alone limbs—but the ancestors of snakes had both forelimbs and hindlimbs.

According to cladistics, hagfishes and lampreys are separate and distinct groups within the chordates (Chordata) [*see illustration on page 113*]. At some point in their lives, all chordates display the following characteristics: a hollow, dorsal nerve cord; a notochord, situated immediately below the nerve cord; gill slits; and a segmentally muscled tail that extends past the anus. Hagfishes are considered the most primitive living craniates. Lampreys also have a cranium, but unlike hagfishes, they also have segments of cartilage to protect their nerve cord. These cartilage segments are the first evolutionary rudiments of a backbone, or vertebral column. Lampreys, therefore, are considered the most primitive living vertebrates.

By comparing fossils with living creatures, biologists can create diagrams called cladograms that display the evolutionary relations among organisms. A cladogram of the chordates suggests that the hagfish diverged from the vertebrate evolutionary line around 530 million years ago. It also reveals that the predecessors of the hagfish never had a bony skeleton but that those of the lamprey did. What is more, the cladogram suggests that all early craniates had a complex protrusible feeding apparatus comparable to that of hagfishes. Early vertebrates, including the distant ancestors of humans, probably shared many other anatomical and physiological characteristics with modern hagfishes. But hagfishes have evolved many unique specializations: their eyes and lateral lines regressed, and they developed slime glands.

Besides their key position on the tree of life, hagfishes are also gaining new respect as members of the complex ecosystem of the ocean bottom. Scientists now know that the animals are more abundant than was once thought. Based on trapping surveys done between 1987 and 1992, my colleagues and I estimated that the inner Gulf of Maine contains population densities of up to 500,000 *M. glutinosa* per square kilometer. W. Waldo Wakefield of Rutgers University, who was then at the Scripps Institution of Oceanography, found comparable densities of *E. deani* off the California coast at depths of between 600 and 800 meters.

We also now recognize the degree of hagfish predation on the populations of other animals that live near the bottom of the ocean. Although

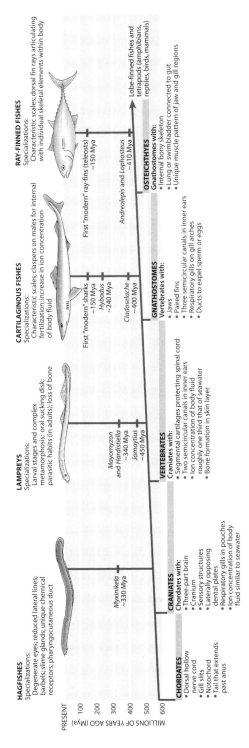

HAGFISHES
Specializations:
Degenerate eyes; reduced lateral lines; barbels; slime glands; unique chemical receptors; pharyngocutaneous duct

LAMPREYS
Specializations:
Larval stages and complex metamorphosis; oral sucking disk; parasitic habits (in adults); loss of bone

CARTILAGINOUS FISHES
Specializations:
Characteristic scales; claspers on males for internal fertilization; increase in ion concentration of body fluid

RAY-FINNED FISHES
Specializations:
Characteristic scales; dorsal fin rays articulating with individual skeletal elements within body

Myxinikela
~330 Mya

Mayomyzon and Hardistiella
~340 Mya

Jamoytius
~450 Mya

First "modern" sharks
~150 Mya

Hybodus
~240 Mya

Cladoselache
~400 Mya

First "modern" ray-fins (teleosts)
~150 Mya

Andreolepis and Lophosteus
~410 Mya

Lobe-finned fishes and tetrapods (amphibians, reptiles, birds, mammals)

CHORDATES
Chordates with:
• Dorsal hollow nerve cord
• Gill slits
• Notochord
• Tail that extends past anus

CRANIATES
Chordates with:
• Three-part brain
• Cranium
• Sensory structures
• Laterally opposing dental plates
• Respiratory gills in pouches
• Ion concentration of body fluid similar to seawater

VERTEBRATES
Craniates with:
• Segmental cartilages protecting spinal cord
• Two semicircular canals in inner ears
• Ion concentration of body fluid roughly one third that of seawater
• Bone formation in skin layer

GNATHOSTOMES
Vertebrates with:
• Jaws
• Paired fins
• Three semicircular canals in inner ears
• Respiratory gills on gill arches
• Ducts to expel sperm or eggs

OSTEICHTHYES
Gnathostomes with:
• Internal bony skeleton
• Lung or swimbladder connected to gut
• Unique muscle pattern of jaw and gill regions

PRESENT
100
200
300
400
500
600
MILLIONS OF YEARS AGO (Mya)

Evolutionary tree, or cladogram, shows that hagfishes are the oldest surviving craniates (animals with a braincase)—predating lampreys, cartilaginous fishes and ray-finned fishes by millions of years. (Fossils are indicated in italics.)

individual hagfishes have extremely low metabolic rates, their energy needs add up. The average number of M. *glutinosa* that inhabit one square kilometer of seafloor (59,700 animals) must consume the caloric equivalent of 18.25 metric tons of shrimp, 11.7 metric tons of sea worms or 9.9 metric tons of fish every year. And that amount would be sufficient only to keep the animals alive at rest; when they swim or burrow, their energy demands increase between four and five times.

Hagfishes also consume discarded so-called bycatch from commercial trawl-fishing fleets and play a central role in recycling the carcasses of dead marine vertebrates, including whales. Craig Smith of the University of Hawaii has found that hagfishes can remove roughly 90 percent of the energy content of small packages of bait sunk to the seafloor at depths of 1,200 meters. But hagfishes are not just important ecologically for their roles as predators or scavengers, they also serve as important prey for a surprising number of marine animals, including codfish, dogfish sharks, octopuses, cormorants, harbor porpoises, harbor seals, elephant seals and some species of dolphins.

THE SLIME HAG TRADE

In many locales around the world, hagfishes have become the focus of a large and flourishing commercial fishery. Since the 1960s there has been a booming trade in leather goods produced from tanned hagfish skin. These products, which are manufactured primarily in South Korea, are sold as "eel-skin" (presumably because consumers would be less likely to pay high prices for goods labeled "slime-hag hide").

Hagfish skin, which is smooth and slick to the touch, consists of a superficial layer of epidermis overlying a dermis containing multiple, dense layers of collagen fibers. In the leather preparation process, the epidermis is removed, and the treated dermis is used to produce designer handbags, shoes, wallets, purses, briefcases and so forth. Removing the skin is relatively easy because it is attached to underlying muscles only along the dorsal midline and along the ventral surface at the level of the slime glands. Thus, the leather produced by one hagfish is a long strip, with a wrinkled band down the midline marking the site of the dorsal attachment to muscle.

The demand for suitable skins has supported commercial hagfish ventures around the Pacific Rim and in the western North Atlantic. The strip of skin must be above a minimum width (roughly five centimeters) but must not be too thick. This combination eliminates many species from the fishery: some are too small or too slender, others too large and too thick-skinned. The collection method is very low tech: multiple traps baited with anything from herring to kitchen scraps are set along a line on the sea bottom and left overnight. The traps can be 19-liter pails with lids or 190-liter barrels with small holes in the sides; once inside, most hagfishes become trapped in the bait and their own slime. In previously unfished areas, more than 100 hagfishes have been found to enter a given trap during its first hour on the bottom.

Unfortunately, the demand for hagfish skin has depleted the populations of many species because the trapping rate has far exceeded their rate of reproduction. And as each species becomes less abundant, fishers target others. Over the past three decades, fishers have exploited *Paramyxine atami, E. burgeri, E. okinoseanus* and *M. garmani* in the western North Pacific, *E. stouti* and *E. deani* off the Pacific coast of North America, and *M. glutinosa* in the Gulf of Maine.

In the region of New England, annual landings of hagfishes went from virtually zero in 1991 to roughly 1,950 metric tons in 1996. Over that five-year span, roughly 50 million hagfishes were processed and shipped overseas. Fishers discard hagfishes that are shorter than 500 millimeters—the minimum length suitable for leather—and these usually die when released into the comparatively warm surface waters. Accordingly, the actual impact of fishing on hagfish populations is far greater than indicated by the landings alone. By 1996 there were signs that hagfish fisheries were in trouble; recent declines in landings, average size and catch per trap suggest that the trouble is serious.

This state of affairs is not likely to improve, because almost everywhere hagfishes are classified as "underutilized species," and their exploitation is usually permitted without efforts to regulate the effect on hagfish populations. Hagfish trapping was considered a growth industry in New England, for example, when other fisheries were nearing a state of collapse.

Although hagfishes are more plentiful than once thought, we do not yet know enough about them to manage a sustainable hagfish fishery. In the meantime, we should take simple steps—such as requiring holes in commercial traps through which small, young hagfishes could escape—to reduce the impact on hagfish populations. When we drastically reduce the number of any species—even the lowly (and to some, loathsome) hagfish—we are performing an ecosystem experiment on a grand scale. As usual, we cannot yet begin to predict the eventual results.

FURTHER READING

THE BIOLOGY OF MYXINE. Edited by Alf Brodal and Ragnar Fänge. Universitetsforlaget, Oslo, 1963.

THE HAGFISH. David Jensen in *Scientific American,* Vol. 214, No. 2, pages 82–90; February 1966.

THE HAGFISHERY OF JAPAN. Aubrey Gorbman et al. in *Fisheries,* Vol. 15, No. 4, pages 12–18; July 1990.

THE BIOLOGY OF HAGFISHES. Edited by J. M. Jørgensen, J. P. Lomholt, R. E. Weber and H. Malte. Chapman & Hall, 1998.

Why Are Reef Fish So Colorful?

JUSTIN MARSHALL

ORIGINALLY PUBLISHED IN *SCIENTIFIC AMERICAN PRESENTS*, FALL 1998

Strangely enough, I became curious about the colors of fish not while diving in the crystal-clear waters of Australia's Great Barrier Reef, surrounded by countless incredibly colorful fish. On the contrary: I was in the murky, turbid waters of Heron Island's Coral Cay Lagoon, near the southeastern edge of the reef, close to Shark Bay.

Sitting slightly apprehensively at a depth of only two meters, I was trying to catch fish in a hand net. Suddenly I became dimly aware of hundreds of little black dots shooting past me almost at the limits of my vision in the silty water. Sucking air through my dive regulator and pondering this strange event, I was stunned to realize the black dots were the eyes of an enormous school of kyphosids swimming past on their way to the reef edge. The bodies of these fish, which are also known as drummers, are about 30 centimeters (nearly 12 inches) long and are a silvery-blue color. When vertical in water, they merged perfectly with the dim, blue light pervading the lagoon.

Here was a wonderful example of camouflage underwater. I was humbled by my ineptitude as a predator—I allowed literally tons of fish to pass within a meter or two of me and my net before I even realized what they were. As a marine biologist interested in vision in the sea, however, I immediately thought of several questions. How is the skin of drummers so well adapted to merge with the sea? What is it about the visual capabilities of the fish that prey on drummers that enables them, presumably, to see the drummers while mine was so ineffective?

I noticed that many of the fish and other reef creatures that the school had by now joined were boldly colorful, their bright patterns making them pop out from the background—and also, it would seem, making them an obvious meal. I wondered how the environment of the coral reef could have given rise to the virtually invisible drummer and frogfish as well as the highly conspicuous angelfish and butterfly fish.

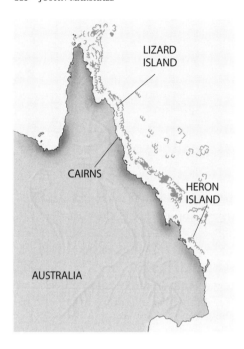

EXTREME BIODIVERSITY

It is such questions that occupy me on field trips to the University of Queensland's research stations on Heron and Lizard islands. These two islands are at either end of the 2,300-kilometer expanse of the Great Barrier Reef (*map*), which is by far the largest reef system in the world and rightfully one of its seven natural wonders. The huge expanse is a living area of 200,000 square kilometers consisting of some 3,000 small reefs that include more than 400 species of hard and soft corals. For comparison, a typical Caribbean reef might be tens of kilometers long and have perhaps 40 kinds of hard and soft corals.

Like terrestrial rain forests, coral reefs are isolated enclaves that are important for their extreme biodiversity. In this respect, too, the Great Barrier Reef is superlative: it is home to around 1,500 species of fish. This huge variety is all the more surprising in light of the relatively young age of the reef. It began to form 12 to 18 million years ago and in some places is only two million years old; reefs of the western Atlantic and central Pacific formed 25 million years ago.

The diversity of colored fish and invertebrates on the Great Barrier Reef is truly awe-inspiring. Yet the color patterns exhibited by these inhabitants did not evolve for human eyes. The brilliant blue spots of the semi-circle angelfish, or the contrasting yellow and blue fins of the yellowtail coris wrasse, are a vital component of the survival strategies of these species on the reef.

To understand this role of color and appearance requires some understanding of survival on the reef and also of the optics of the undersea realm. At its most fundamental level, survival for any animal species demands three things: eating, not being eaten and reproducing. Unfortunately for sea creatures, the demands on appearance imposed by the first two of these survival requirements conflict with those of the third.

A good way to avoid being eaten, or, indeed, to lurk undetected while waiting for prey to swim by, is to be camouflaged to match the background (the scientific term is "cryptic"). Masters of camouflage include the frogfish and the school of kyphosids I saw in the Heron Island lagoon. But to attract a mate, chase rivals away or provide other warnings, bright colors that are easily seen underwater may be the order of the day. In the blue waters of the reef, the colors yellow and blue travel the farthest, so many reef animals have evolved bodily patterns of yellow and blue in striped or spotted combinations. Because yellow and blue are also widely separated in the spectrum, they offer strong contrast underwater.

Just what does a butterfly fish look like to another butterfly fish? How does a drummer appear to a shark? It is this goal to understand color vision and its evolution from the point of view of the animals themselves that my colleagues and I at the University of Queensland's Vision, Touch and Hearing Research Center are striving toward at present. Our research has revolved around three critical questions: One, what are the animals' visual capabilities? To explore this matter, we are carrying out experiments in which we are quantifying colors nonsubjectively, using the world's first underwater spot-reflectance spectroradiometer. Two, what are the light and surroundings like in the habitat where these creatures live? Experienced divers know that seawater is so blue that all red light is absorbed within 20 meters of the surface; a bright-red fish at this depth therefore appears black. And three, under what circumstances, and to what other creatures, do fish show off their color patterns? Clearly, displaying bright colors to impress a potential mate would be unwise when visually guided predators are lurking nearby.

Compared with some species of fish and other creatures, humans are relatively color-blind. People have three color receptors in their eyes: the blue-, green- and red-sensitive cones. Some reef fish (and indeed amphibians, reptiles, birds and insects) possess four or more. The record is currently held by the mantis shrimp (a stomatopod), a reef dweller whose eyes have 12 color receptors. With these additional receptors, the animals can see the region of the near ultraviolet, with wavelengths between about 350 and 400 nanometers (humans cannot see wavelengths shorter than about 380 nanometers). Also, they can see in greater detail some of the colors humans see.

Such impressive visual capabilities might seem to be unnecessary on the reef, where so many creatures have evolved bold patterns that emit strong visual signals. Alternatively, it may seem incredible that these brightly colored fish manage to survive with markings so striking that they would seem to attract the attention of even weak-eyed predators.

Could it be that coral reefs are colorful, and therefore that colorful animals fit in and may even be camouflaged? Logical though it may seem, the notion does not hold up to scrutiny. A reef stripped of its fish and other mobile life-forms is actually relatively monochromatic. Most of the corals are brown or green, their colorful splendor coming out only at night when the polyps open or under the falsely bright illumination of the camera strobe or video light.

Another possible explanation revolves around disruptive coloration, a principle first described in detail in the 1940s and subsequently used for military camouflage. The central idea is the use of large, bold patterns of contrasting colors that make an object blend in when viewed against an equally variable, contrasting background. The light and dark branches, pockets and shafts of light on a reef provide just such a background.

Good examples of disruptive camouflage on land are the striking patterns of some snakes or the stripes of a zebra. These colorations, when viewed against the kinds of highly patterned backgrounds common in the animals' natural habitats, aid camouflage or at least make it difficult to see where the animal begins and ends. For example, zebras—like many boldly colored fish—group together for protection. In these groupings, the context against which predators see individual patterns and colors is not, typically, a natural background but rather the school or herd itself, enabling one animal to become lost in the swirling mass of its neighbors.

Complicating matters is the fact that most reef fish are capable of changing colors to some degree. Some, notably the triggerfish and goatfish, can do so at nearly the speed and complexity of chameleons. In other fish, color changes may take several seconds, may be associated with night and day, or may occur during maturation. Parrot fish change color in association with a sex change, a drab female in a harem changing into a gaudy dominant male if the resident male is lost. Changes are also known to occur with "mood"—for example, during conflicts or flight from a predator. Although one can guess at the causes behind these and other color changes, at present almost no convincing hypotheses explain their function.

Parts of fish may be disguised by a pattern, such as the dark stripes that run near the eyes of the Moorish idol. Similarly, colorations may make it difficult for a predator to determine which end is the head and which is the tail. Many species of butterfly fish, for example, have a black dot on either side of the body near the caudal fin; these dots are easily mistaken for eyes. In patterns on other fish species, blocks of blue match the blue of the ocean.

The effectiveness of highly contrasting body stripes, spots and blotches as a means of reef camouflage can be fully appreciated only under natural illumination. Yet few people get to see fish this way: often reef creatures are viewed in photographs, their colors set ablaze by the flash of a strobe and against a background that is nothing more than a featureless, dark field. Lit up in this manner, the fish are being seen as they are when they are "displaying." Fish sometimes position themselves in shafts of sunlight to reveal the full splendor of their colors to a rival or potential mate. At other times, even the multicolored harlequin tuskfish or iridescent blue angelfish disappear under the dim, highly textured illumination of the coral ledges where they spend much of their time.

Also, just as birds will puff out and spread their feathers in dramatic displays, some reef fish will erect highly colored fins or reveal bright patterns on frontal head areas or even inside the mouth. The positioning of the fish relative to the viewer is obviously critical here; frontal regions are obscured when the fish is seen from the side, whereas the erect-fin display-ers such as the butterfly fish will intentionally turn sideways to present a broadside of color to a rival or mate.

Bright colors can also warn of toxicity. Boxfish, blue-ringed octopuses and nudibranchs are all known for such aposematic displays, in which, again, yellow and blue are a common theme underwater. In contrast to the furtive behavior of animals that are disruptively camouflaged by their bright patterns, however, aposematic displays are generally accompanied by bold and indiscreet behavior. As an interesting side note, evolution has produced aposematic animals unable to see their own beautiful colors. This is the case for nudibranchs and for the blue-ringed octopus. Both these invertebrates lack the retinal features necessary to see colors, indi-cating that their bright patterns evolved solely in response to their preda-tors' much more capable visual systems.

Whether for display or camouflage, the visual signals emanating from reef fish all depend strongly on contrast, and this aspect appears to have evolved with unexpected elegance. As noted earlier, yellow and blue are an effective combination, with peaks in different parts of the spectrum. The two colors are said to be complementary—exhibiting a high degree of contrast—because of this spacing of their spectral peaks.

The spectral characteristics of the colors of several other reef fish are even more complex, with three rather than two peaks. Where this is the case—in the facial displays of wrasses and parrot fish, for example—adjacent colors are also complementary, with each color having a spectral peak that fits neatly into the trough of the adjacent color. Three years ago we began to suspect that some of these exquisitely tuned combinations, not obviously contrasting to us because of the relatively limited color perception of humans, provide particularly strong visual signals to certain fish species.

As noted, color vision in some reef fish and other animals may be based on four photoreceptor types rather than three, as in humans. Because the additional sensitivity afforded by the extra photoreceptor is often in the ultraviolet, we became interested in the possibility that the visual signals sent by a select number of reef fish encompass the ultraviolet as well as the colors visible to humans.

Using our spot-reflectance spectroradiometer, we found this indeed to be the case. The advantage of this device is that it can "see" colors we cannot, including both the near-ultraviolet and the near-infrared regions of the electromagnetic spectrum (with wavelengths of 300 to 400 nanometers and 700 to 800 nanometers, respectively). As a result, we can begin to understand how color patterns have evolved for animals that see these colors. Our work has involved trying to establish what the various reef fish can see. Our most recent results indicate that in adult life, a relatively small proportion of reef fish see the near ultraviolet. As with aposematic coloration, however, it is becoming clear that animal colors are not necessarily correlated with their own visual systems.

Although the exact function of this possibly "secret waveband" remains a mystery, ultraviolet is in theory a good color for local signaling. The fact that ultraviolet is highly scattered and attenuated by water means, for example, that the visual signals of a sexual display could be sent to a nearby potential mate—and that the signal would degrade to invisibility over the longer distances at which predators might lurk.

There are many related issues about which we know little. For instance, color vision changes substantially during the life spans of reef fish. For example, it appears that the eyes of reef fish larvae do not block ultraviolet, and yet most of the adults of these species cannot see this part of the spectrum. We know that the change is to accommodate the demands of a new mode of life—the emergence from the plankton, where all fish begin life. So far, however, the details of this vision change are known for only two of the 1,500 species on the Great Barrier Reef.

This is just one of the mysteries that leave vast gaps in our knowledge. We still have only fragmentary ideas about what the colors of a reef mean to its inhabitants, making each visit to this world of secret color communication an endeavor as tantalizing as it is beautiful.

FISHERIES

Counting the Last Fish

DANIEL PAULY AND REG WATSON

ORIGINALLY PUBLISHED IN JULY 2003

Georges Bank—the patch of relatively shallow ocean just off the coast of Nova Scotia, Canada—used to teem with fish. Writings from the 17th century record that boats were often surrounded by huge schools of cod, salmon, striped bass and sturgeon. Today it is a very different story. Trawlers trailing dredges the size of football fields have literally scraped the bottom clean, harvesting an entire ecosystem—including supporting substrates such as sponges—along with the catch of the day. Farther up the water column, longlines and drift nets are snagging the last sharks, swordfish and tuna. The hauls of these commercially desirable species are dwindling, and the sizes of individual fish being taken are getting smaller; a large number are even captured before they have time to mature. The phenomenon is not restricted to the North Atlantic but is occurring across the globe.

Many people are under the mistaken impression that pollution is responsible for declines in marine species. Others may find it hard to believe that a shortage of desirable food fish even exists, because they still notice piles of Chilean sea bass and tuna fillets in their local fish markets. Why is commercial fishing seen as having little if any effect on the species that are being fished? We suspect that this perception persists from another age, when fishing was a matter of wresting sustenance from a hostile sea using tiny boats and simple gear.

Our recent studies demonstrate that we can no longer think of the sea as a bounteous provider whose mysterious depths contain an inexhaustible resource. Over the past several years we have gathered and analyzed data on the world's fisheries, compiling the first comprehensive look at the state of the marine food resource. We have found that some countries, particularly China, have overreported their catches, obscuring a downward trend in fish caught worldwide. In general, fishers must work farther offshore and at greater depths in an effort to keep up with the catches of yesteryear and to try to meet the burgeoning demand for fish.

We contend that overfishing and the fishing of these distant stocks are unsustainable practices and are causing the depletion of important species. But it is not too late to implement policies to protect the world's fisheries for future generations.

THE LAW OF THE SEA

Explaining how the sea got into its current state requires relating a bit of history. The ocean used to be a free-for-all, with fleets flying the flags of various countries competing for fish thousands of miles from home. In 1982 the United Nations adopted the Convention on the Law of the Sea, which allows countries bordering the ocean to claim exclusive economic zones reaching 200 nautical miles into open waters. These areas include the highly productive continental shelves of roughly 200 meters in depth where most fish live out their lives.

The convention ended decades—and, in some instances, even centuries—of fighting over coastal fishing grounds, but it placed the responsibility for managing marine fisheries squarely on maritime countries. Unfortunately, we cannot point to any example of a nation that has stepped up to its duties in this regard.

The U.S. and Canadian governments have subsidized the growth of domestic fishing fleets to supplant those of now excluded foreign countries. Canada, for instance, built new offshore fleets to replace those of foreign nations pushed out by the convention, effectively substituting foreign boats with even larger fleets of more modern vessels that fish year-round on the same stocks that the domestic, inshore fleet was already targeting. In an effort to ensure that there is no opportunity for foreign fleets to fish the excess allotment—as provided for in the convention—these nations have also begun to fish more extensively than they would have otherwise. And some states, such as those in West Africa, have been pressured by others to accept agreements that allow foreign fleets to fish their waters, as sanctioned by the convention. The end result has been more fishing than ever, because foreign fleets have no incentive to preserve local marine resources long-term—and, in fact, are subsidized by their own countries to garner as much fish as they can.

The expansion made possible by the Convention on the Law of the Sea and technological improvements in commercial fishing gear (such as acoustic fish finders) temporarily boosted fish catches. But by the late 1980s the upward trend began to reverse, despite overreporting by China, which, in order to meet politically driven "productivity increases," was stating that it was taking nearly twice the amount of fish that it actually was.

In 2001 we presented a statistical model that allowed us to examine where catches differed significantly from those taken from similarly productive waters at the same depths and latitudes elsewhere in the world. The figures from Chinese waters—about 1 percent of the world's oceans—were much higher than predicted, accounting for more than 40 percent of the deviations from the statistical model. When we readjusted the worldwide fisheries data for China's misrepresentations, we concluded that world fish landings have been declining slowly since the late 1980s, by about 700,000 metric tons a year. China's overreporting skewed global fisheries statistics so significantly because of the country's large size and the degree of its overreporting. Other nations also submit inaccurate fisheries statistics—with a few overreporting their catches and most underreporting them—but those numbers tend to cancel one another out.

Nations gather statistics on fish landings in a variety of ways, including surveys, censuses and logbooks. In some countries, such as China, these data are forwarded to regional offices and on up through the government hierarchy until they arrive at the national offices. At each step, officials may manipulate the statistics to meet mandatory production targets. Other countries have systems for cross-checking the fish landings against import/export data and information on local consumption.

The most persuasive evidence, in our opinion, that fishing is wreaking havoc on marine ecosystems is the phenomenon that one of us (Pauly) has dubbed "fishing down the food web." This describes what occurs when fishers deplete large predator fish at the top of the food chain, such as tuna and swordfish, until they become rare, and then begin to target smaller species that would usually be eaten by the large fish [*see illustration on page 130*].

FISHING DOWN

The position a particular animal occupies in the strata of a food web is determined by its size, the anatomy of its mouthparts and its feeding preferences. The various layers of the food web, called trophic levels, are ranked according to how many steps they are removed from the primary producers at the base of the web, which generally consists of phytoplanktonic algae. These microscopic organisms are assigned a trophic level (TL) of 1.

Phytoplankton are grazed mostly by small zooplankton—mainly tiny crustaceans of between 0.5 and two millimeters in size, both of which thus have a TL of 2. (This size hierarchy stands in stark contrast to terrestrial food chains, in which herbivores are often very large; consider moose or elephants, for instance.) TL 3 consists of small fishes between 20 and

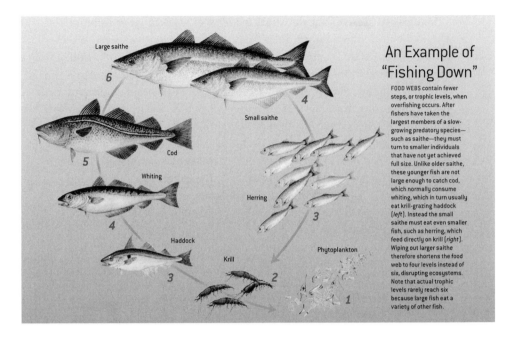

An Example of "Fishing Down"

FOOD WEBS contain fewer steps, or trophic levels, when overfishing occurs. After fishers have taken the largest members of a slow-growing predatory species—such as saithe—they must turn to smaller individuals that have not yet achieved full size. Unlike older saithe, these younger fish are not large enough to catch cod, which normally consume whiting, which in turn usually eat krill-grazing haddock [*left*]. Instead the small saithe must eat even smaller fish, such as herring, which feed directly on krill [*right*]. Wiping out larger saithe therefore shortens the food web to four levels instead of six, disrupting ecosystems. Note that actual trophic levels rarely reach six because large fish eat a variety of other fish.

50 centimeters in length, such as sardines, herring and anchovies. These small pelagic fishes live in open waters and usually consume a variable mix of phytoplankton and both herbivorous and carnivorous zooplankton. They are caught in enormous quantities by fisheries: 41 million metric tons were landed in 2000, a number that corresponds to 49 percent of the reported global marine fish catch. Most are either destined for human consumption, such as canned sardines, or reduced to fish meal and oil to serve as feed for chickens, pigs and farmed salmon or other carnivorous fish.

The typical table fish—the cod, snapper, tuna and halibut that restaurants serve whole or as steaks or fillets—are predators of the small pelagics and other small fishes and invertebrates; they tend to have a TL of between 3.5 and 4.5. (Their TLs are not whole numbers because they can consume prey on several trophic levels.)

The increased popularity in the U.S. of such fish as nutritious foods has undoubtedly contributed to the decline in their stocks. We suggest that the health and sustainability of fisheries can be assessed by monitoring the trends of average TLs. When those numbers begin to drop, it indicates that fishers are relying on ever smaller fish and that stocks of the larger predatory fish are beginning to collapse.

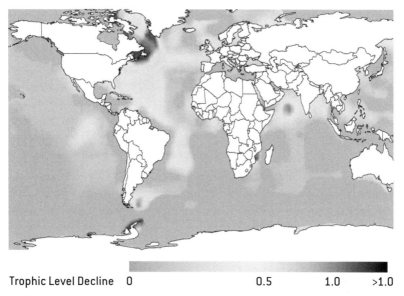

Trophic Level Decline   0   0.5   1.0   >1.0

Overfishing caused the complexity of the food chains in important fisheries to drop by more than one trophic level between the years 1950 and 2000. The open ocean usually has few fish.

In 1998 we presented the first evidence that "fishing down" was already occurring in some fishing grounds, particularly in the North Atlantic, off the Patagonian coast of South America and nearby Antarctica, in the Arabian Sea, and around parts of Africa and Australia. These areas experienced TL declines of 1 or greater between 1950 and 2000, according to our calculations [*see map above*]. Off the west coast of Newfoundland, for instance, the average TL went from a maximum of 3.65 in 1957 to 2.6 in 2000. Average sizes of fish landed in those regions dropped by one meter during that period.

Our conclusions are based on an analysis of the global database of marine fish landings that is created and maintained by the U.N. Food and Agriculture Organization, which is in turn derived from data provided by member countries. Because this data set has problems—such as overreporting and the lumping of various species into a category called "mixed"—we had to incorporate information on the global distribution of fishes from FishBase, the online encyclopedia of fishes pioneered by Pauly, as well as information on the fishing patterns and access rights of countries reporting catches.

Research by some other groups—notably those led by Jeremy B. C. Jackson of the Scripps Institution of Oceanography in San Diego and Ransom A.

Myers of Dalhousie University in Halifax—suggests that our results, dire as they might seem, in fact underestimate the seriousness of the effects that marine fisheries have on their underlying resources. Jackson and his colleagues have shown that massive declines in populations of marine mammals, turtles and large fishes occurred along all coastlines where people lived long before the post–World War II period we examined. The extent of these depletions was not recognized until recently because biologists did not consult historians or collaborate with archaeologists, who study evidence of fish consumption in middens (ancient trash dumps).

Myers and his co-workers used data from a wide range of fisheries throughout the world to demonstrate that industrial fleets generally take only a few decades to reduce the biomass of a previously unfished stock by a factor of 10. Because it often takes much longer for a regulatory regime to be established to manage a marine resource, the sustainability levels set are most likely to be based on numbers that already reflect population declines. Myers's group documents this process particularly well for the Japanese longline fishery, which in 1952 burst out of the small area around Japan—to which it was confined until the end of the Korean War—and expanded across the Pacific and into the Atlantic and Indian oceans. The expansion decimated tuna populations worldwide. Indeed, Myers and his colleague Boris Worm recently reported that the world's oceans have lost 90 percent of large predatory fish.

CHANGING THE FUTURE

What can be done? Many believe that fish farming will relieve the pressure on stocks, but it can do so only if the farmed organisms do not consume fish meal. (Mussels, clams and tilapia, an herbivorous fish, can be farmed without fish meal.) When fish are fed fish meal, as in the case of salmon and various carnivores, farming makes the problem worse, turning small pelagics—including fish that are otherwise perfectly fit for human consumption, such as herring, sardines, anchovies and mackerels—into animal fodder. In fact, salmon farms consume more fish than they produce: it can take three pounds of fish meal to yield one pound of salmon.

One approach to resolving the difficulties now besetting the world's fisheries is ecosystem-based management, which would seek to maintain—or, where necessary, reestablish—the structure and function of the ecosystems within which fisheries are embedded. This would involve considering the food requirements of key species in ecosystems (notably those of marine mammals), phasing out fishing gear that destroys the sea bot-

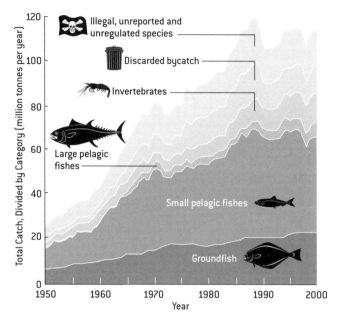

Amount of fish landed has more than quintupled over the past 50 years. As the world's population has grown, commercial fishing technology has advanced, and demand for fish in some countries has surged.

tom, and implementing marine reserves, or "no-take zones," to mitigate the effects of fishing. Such strategies are compatible with the set of reforms that have been proposed for years by various fisheries scientists and economists: radically reducing global fleet capacity; abolishing government subsidies that keep otherwise unprofitable fishing fleets afloat; and strictly enforcing restrictions on gear that harm habitats or that capture "bycatch," species that will ultimately be thrown away.

Creating no-take zones will be key to preserving the world's fisheries. Some refuges should be close to shore, to protect coastal species; others must be large and offshore, to shield oceanic fishes. No-take zones now exist, but they are small and scattered. Indeed, the total area protected from any form of fishing constitutes a mere 0.01 percent of the ocean surface. Reserves are now viewed by fishers—and even by governments—as necessary concessions to conservationist pressure, but they must become management tools for protecting exploited species from overfishing.

A major goal should be to conserve species that once maintained themselves at deeper depths and farther offshore, before fishers developed improved gear for going after them. This type of fishing is similar to a nonrenewable mining operation because fishes are very vulnerable, typically long-lived, and have very low productivity in the dark, cold depths. These measures would enable fisheries, for the first time, to become sustainable.

FURTHER READING

EFFECT OF AQUACULTURE ON WORLD FISH SUPPLIES. Rosamond L. Naylor, Rebecca J. Goldburg, Jurgenne H. Primavera, Nils Kautsky, Malcolm C. M. Beveridge, Jason Clay, Carl Folke, Jane Lubchenco, Harold Mooney and Max Troell in *Nature,* Vol. 405, pages 1017–1024; June 29, 2000.

HISTORICAL OVERFISHING AND THE RECENT COLLAPSE OF COASTAL ECOSYSTEMS. Jeremy B. C. Jackson et al. in *Science,* Vol. 293, pages 629–638; July 27, 2001.

SYSTEMATIC DISTORTION IN WORLD FISHERIES CATCH TRENDS. Reg Watson and Daniel Pauly in *Nature,* Vol. 414, pages 534–536; November 29, 2001.

IN A PERFECT OCEAN: THE STATE OF FISHERIES AND ECOSYSTEMS IN THE NORTH ATLANTIC OCEAN. Daniel Pauly and Jay Maclean. Island Press, 2003.

RAPID WORLDWIDE DEPLETION OF PREDATORY FISH COMMUNITIES. Ransom A. Myers and Boris Worm in *Nature,* Vol. 423, pages 280–283; May 15, 2003.

More information on the state of world fisheries can be found on the Web sites of the Sea Around Us Project at www.saup.fisheries.ubc.ca and of FishBase at www.fishbase.org

The World's Imperiled Fish

CARL SAFINA

ORIGINALLY PUBLISHED IN NOVEMBER 1995

The 19th-century naturalist Jean-Baptiste de Lamarck is well known for his theory of the inheritance of acquired characteristics, but he is less remembered for his views on marine fisheries. In pondering the subject, he wrote, "Animals living in . . . the sea waters . . . are protected from the destruction of their species by man. Their multiplication is so rapid and their means of evading pursuit or traps are so great, that there is no likelihood of his being able to destroy the entire species of any of these animals." Lamarck was also wrong about evolution.

One can forgive Lamarck for his inability to imagine that humans might catch fish faster than these creatures could reproduce. But many people—including those in professions focused entirely on fisheries—have committed the same error of thinking. Their mistakes have reduced numerous fish populations to extremely low levels, destabilized marine ecosystems and impoverished many coastal communities. Ironically, the drive for short-term profits has cost billions of dollars to businesses and taxpayers, and it has threatened the food security of developing countries around the world. The fundamental folly underlying the current decline has been a widespread failure to recognize that fish are wildlife—the only wildlife still hunted on a large scale.

Because wild fish regenerate at rates determined by nature, attempts to increase their supply to the marketplace must eventually run into limits. That threshold seems to have been passed in all parts of the Atlantic, Mediterranean and Pacific: these regions each show dwindling catches. Worldwide, the extraction of wild fish has seemingly stagnated at about 84 million metric tons.

In some areas where the catches peaked as long ago as the early 1970s, landings have since decreased by more than 50 percent. Even more disturbingly, some of the world's greatest fishing grounds, including the Grand Banks and Georges Bank of eastern North America, are now essentially closed following their collapse. The formerly dominant fauna have been

135

Major Fishing Regions of the World: Changes in Catch

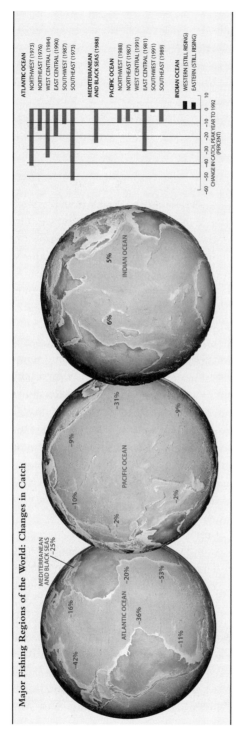

ATLANTIC OCEAN
NORTHWEST (1973)
NORTHEAST (1976)
WEST CENTRAL (1984)
EAST CENTRAL (1990)
SOUTHWEST (1987)
SOUTHEAST (1973)

MEDITERRANEAN
AND BLACK SEAS (1988)

PACIFIC OCEAN
NORTHWEST (1988)
NORTHEAST (1987)
WEST CENTRAL (1991)
EAST CENTRAL (1981)
SOUTHWEST (1991)
SOUTHEAST (1989)

INDIAN OCEAN
WESTERN (STILL RISING)
EASTERN (STILL RISING)

-60 -50 -40 -30 -20 -10 0 10
CHANGE IN CATCH, PEAK YEAR TO 1992
(PERCENT)

MEDITERRANEAN
AND BLACK SEAS -25%

ATLANTIC OCEAN -20%

-16%

-42%

-36%

-53%

-11%

PACIFIC OCEAN

-31%

-9%

-10%

-9%

-2%

-2%

INDIAN OCEAN

5%

6%

Regional takes of fish have fallen in most areas of the globe, having reached their peak values anywhere from seven to 25 years ago. (The year of the peak catch is shown in parentheses.) Only in the Indian Ocean region, where modern methods of mechanized fishing are just now taking hold, have marine catches been on the increase. (Black indicates average annual growth between 1988 and 1992.)

reduced to a tiny fraction of their previous abundance and effectively rendered commercially extinct in those areas.

Recognizing that a basic shift has occurred, the members of the United Nations's Food and Agriculture Organization (a body that encouraged the expansion of large-scale industrial fishing during the 1980s) recently concluded that the operation of the world's fisheries cannot be sustained. They now acknowledge that substantial damage has already been done to the marine environment and to the many economies that depend on this crucial natural resource.

Such sobering assessments are echoed in the U.S. by the National Academy of Sciences. It reported in 1995 that human actions have caused drastic reductions in many of the preferred species of edible fish and that changes induced in composition and abundance of marine animals and plants are extensive enough to endanger the functioning of marine ecosystems. Although the scientists involved in that study noted that fishing constitutes just one of many human activities that threaten the oceans, they ranked it as the most serious.

Indeed, the environmental problems facing the seas are in some ways more pressing than those on land. Daniel Pauly of the Fisheries Center at the University of British Columbia and Villy Christensen of the International Center for Living Aquatic Resources Management in Manila have pointed out that the vast majority of shallow continental shelves have been scarred by fishing, whereas large untouched tracts of rain forest still exist. For those who work with living marine resources, the damage is not at all subtle. Vaughn C. Anthony, a scientist formerly with the National Marine Fisheries Service, has said simply: "Any dumb fool knows there's no fish around."

A WAR ON FISHES

How did this collapse happen? An explosion of fishing technologies occurred during the 1950s and 1960s. During that time, fishers adapted various military technologies to hunting on the high seas. Radar allowed boats to navigate in total fog, and sonar made it possible to detect schools of fish deep under the oceans' opaque blanket. Electronic navigation aids such as LORAN (Long-Range Navigation) and satellite positioning systems turned the trackless sea into a grid so that vessels could return to within 15 meters of a chosen location, such as sites where fish gathered and bred. Ships can now receive satellite weather maps of water-temperature fronts, indicating where fish will be traveling. Some vessels work in concert with aircraft used to spot fish.

Many industrial fishing vessels are floating factories deploying gear of enormous proportions: 129 kilometers of submerged longlines with thousands of baited hooks, bag-shaped trawl nets large enough to engulf 12 jumbo jetliners and 64-kilometer-long drift nets (still in use by some countries). Pressure from industrial fishing is so intense that 80 to 90 percent of the fish in some populations are removed every year.

For the past two decades, the fishing industry has had increasingly to face the result of extracting fish faster than these populations could reproduce. Fishers have countered loss of preferred fish by switching to species of lesser value, usually those positioned lower in the food web—a practice that robs larger fish, marine mammals and seabirds of food. During the 1980s, five of the less desirable species made up nearly 30 percent of the world fish catch but accounted for only 6 percent of its monetary value. Now there are virtually no untapped marine fish that can be exploited economically.

With the decline of so many species, some people have turned to raising fish to make up for the shortfall. Aquaculture has doubled its output in the past decade, increasing by about 10 million metric tons since 1985. The practice now provides more freshwater fish than wild fisheries do. Saltwater salmon farming also rivals the wild catch, and about half the shrimp now sold are raised in ponds. Overall, aquaculture supplies one third of the fish eaten by people.

Unfortunately, the development of aquaculture has not reduced the pressure on wild populations. Strangely, it may do the opposite. Shrimp farming has created a demand for otherwise worthless catch that can be used as feed. In some countries, shrimp farmers are now investing in trawl nets with fine mesh to catch everything they can for shrimp food, a practice known as biomass fishing. Many of the catch are juveniles of valuable species, and so these fish never have the opportunity to reproduce.

Fish farms can hurt wild populations because the construction of pens along the coast often requires cutting down mangroves; the submerged roots of these salt-tolerant trees provide a natural nursery for shrimp and fish. Peter Weber of the Worldwatch Institute reports that aquaculture is one of the major reasons that half the world's mangroves have been destroyed. Aquaculture also threatens marine fish because some of its most valuable products, such as groupers, milkfish and eels, cannot be bred in captivity and are raised from newly hatched fish caught in the wild. The constant loss of young fry then leads these species even further into decline.

Aquaculture also proves a poor replacement for fishing because it requires substantial investment, land ownership and large amounts of clean water. Most of the people living on the crowded coasts of the world lack all these resources. Aquaculture as carried out in many undeveloped nations often produces only shrimp and expensive types of fish for export to richer countries, leaving most of the locals to struggle for their own needs with the oceans' declining resources.

MADHOUSE ECONOMICS

If the situation is so dire, why are fish so available and, in most developed nations, affordable? Seafood prices have, in fact, risen faster than those for chicken, pork or beef, and the lower cost of these foods tends to constrain the price of fish—people would turn to other meats if the expense of seafood far surpassed them.

Further price increases will also be slowed by imports, by overfishing to keep supplies high (until they crash) and by aquaculture. For instance, the construction of shrimp farms that followed the decline of many wild populations has kept prices in check.

So to some extent, the economic law of supply and demand controls the cost of fish. But no law says fisheries need to be profitable. To catch $70 billion worth of fish, the fishing industry recently incurred costs totaling $124 billion annually. Subsidies fill much of the $54 billion in deficits. These artificial supports include fuel-tax exemptions, price controls, low-interest loans and outright grants for gear or infrastructure. Such massive subsidies arise from the efforts of many governments to preserve employment despite the self-destruction of so many fisheries.

These incentives have for many years enticed investors to finance more fishing ships than the seas' resources could possibly support. Between 1970 and 1990, the world's industrial fishing fleet grew at twice the rate of the global catch, fully doubling in the total tonnage of vessels and in number. This armada finally achieved twice the capacity needed to extract what the oceans could sustainably produce. Economists and managers refer to this situation as overcapitalization. Curiously, fishers would have been able to catch as much with no new vessels at all. One U.S. study found that the annual profits of the yellowtail flounder fishery could increase from zero to $6 million by removing more than 100 boats.

Because this excessive capacity rapidly depletes the amount of fish available, profitability often plummets, reducing the value of ships on the market. Unable to sell their chief asset without major financial loss, owners of these vessels are forced to keep fishing to repay their loans and are caught in an economic trap. They often exercise substantial political pressure so that government regulators will not reduce allowable takes. This common pattern has become widely recognized. Even the U.N. now acknowledges that by enticing too many participants, high levels of subsidy ultimately generate severe economic and environmental hardship.

A WORLD GROWING HUNGRIER

While the catch of wild marine fish declines, the number of people in the world increases every year by about 100 million, an amount equal to the current population of Mexico. Maintaining the present rate of consumption in the face of such growth will require that by 2010 approximately 19 million additional metric tons of seafood become available every year. To achieve this level, aquaculture would have to double in the next 15 years, and wild fish populations would have to be restored to allow higher sustainable catches.

Technical innovations may also help produce human food from species currently used to feed livestock. But even if all the fish that now go to these animals—a third of the world catch—were eaten by people, today's average consumption could hold for only about 20 years. Beyond that time, even improved conservation of wild fish would not be able to keep pace with human population growth. The next century will therefore witness the heretofore unthinkable exhaustion of the oceans' natural ability to satisfy humanity's demand for food from the seas.

To manage this limited resource in the best way possible will clearly require a solid understanding of marine biology and ecology. But substantial difficulties will undoubtedly arise in fashioning scientific information into intelligent policies and in translating these regulations into practice. Managers of fisheries as well as policymakers have for the most part ignored the numerous national and international stock assessments done in past years.

Where regulators have set limits, some fishers have not adhered to them. From 1986 to 1992, distant water fleets fishing on the international part of the Grand Banks off the coast of Canada removed 16 times the quotas for cod, flounder and redfish set by the Northwest Atlantic Fisheries Organization. When Canadian officials seized a Spanish fishing boat near the Grand Banks in 1995, they found two sets of logbooks—one recording true operations and one faked for the authorities. They also discovered nets with illegally small mesh and 350 metric tons of juvenile Greenland halibut. None of the fish on board were mature enough to have reproduced. Such selfish disregard for regulations helped to destroy the Grand Banks fishery.

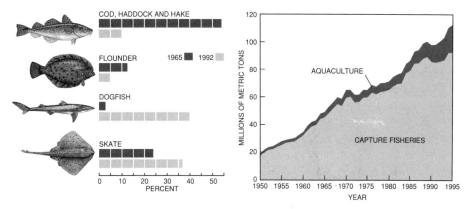

Relative abundance of common fishes in the Gulf of Maine has changed drastically because of over-fishing. The horizontal bars indicate the fraction of the catch made up of each of these species in 1965 as compared with 1992. Fish supplies derived from aquaculture continue to rise steadily, but the total amount available from capture fisheries (which provide the greatest share of the global yield) has entered a period of minimal growth over the past decade.

Although the U.N. reports that about 70 percent of the world's edible fish, crustaceans and mollusks are in urgent need of managed conservation, no country can be viewed as generally successful in fisheries management. International cooperation has been even harder to come by. If a country objects to the restrictions of a particular agreement, it just ignores them.

In 1991, for instance, several countries arranged to reduce their catches of swordfish from the Atlantic; Spain and the U.S. complied with the limitations (set at 15 percent less than 1988 levels), but Japan's catch rose 70 percent, Portugal's landings increased by 120 percent and Canada's take nearly tripled. Norway has decided unilaterally to resume hunting minke whales despite an international moratorium. Japan's hunting of minke whales, ostensibly for scientific purposes, supplies meat that is sold for food and maintains a market that supports illegal whaling around the globe.

INNOCENT BYSTANDERS

In virtually every kind of fishery, people inadvertently capture forms of marine life that collectively are known as bycatch or bykill. In the world's commercial fisheries, one of every four animals taken from the sea is unwanted. Fishers simply discard the remains of these numerous creatures overboard.

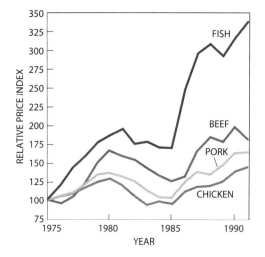

Export prices for fish have exceeded those for beef, chicken and pork by a substantial margin over the past two decades. To facilitate comparison, the price of each meat is scaled to 100 for 1975.

Bycatch involves a variety of marine life, such as species without commercial value and young fish too small to sell. In 1990 high-seas drift nets tangled 42 million animals that were not targeted, including diving seabirds and marine mammals. Such massive losses prompted the U.N. to enact a global ban on large-scale drift nets (those longer than 2.5 kilometers)—although Italy, France and Ireland, among other countries, continue to deploy them.

In some coastal areas, fishing nets set near the sea bottom routinely ensnare small dolphins. Losses to fisheries of several marine mammals—the baiji of eastern Asia, the Mexican vaquita (the smallest type of dolphin known), Hector's dolphins in the New Zealand region and the Mediterranean monk seal—put those species' survival at risk.

Seabirds are also killed when they try to eat the bait attached to fishing lines as these are played out from ships. Rosemary Gales, a research scientist at the Parks and Wildlife Service in Hobart, Tasmania, estimates that in the Southern Hemisphere more than 40,000 albatross are hooked and drowned every year after grabbing at squid used as bait on longlines being set for bluefin tuna. This level of mortality endangers six of the 14 species of these majestic wandering seabirds.

In some fisheries, bykill exceeds target catch. In 1992 in the Bering Sea, fishers discarded 16 million red king crabs, keeping only about three million. Trawling for shrimp produces more bykill than any other type of fishing and accounts for more than a third of the global total. Discarded creatures outnumber shrimp taken by anywhere from 125 to 830 percent.

In the Gulf of Mexico shrimp fishery, 12 million juvenile snappers and 2,800 metric tons of sharks are discarded annually. Worldwide, fishers dispose of about six million sharks every year—half of those caught. And these statistics probably underestimate the magnitude of the waste: much bycatch goes unreported.

There remain, however, some glimmers of hope. The bykill of sea turtles in shrimp trawls had been a constant plague on these creatures in U.S. waters (the National Research Council estimated that up to 55,000 adult turtles die this way every year). But these deaths are being reduced by recently mandated "excluder devices" that shunt the animals out a trap door in the nets.

Perhaps the best-publicized example of bycatch involved up to 400,000 dolphins killed annually by fishers netting Pacific yellowfin tuna. Over three decades since the tuna industry began using huge nets, the eastern spinner dolphin population fell 80 percent, and the numbers of offshore spotted dolphin plummeted by more than 50 percent. These declines led to the use of so-called dolphin-safe methods (begun in 1990) whereby fishers shifted from netting around dolphin schools to netting around logs and other floating objects.

This approach has been highly successful: dolphin kills went down to 4,000 in 1993. Unfortunately, dolphin-safe netting methods are not safe for immature tuna, billfish, turtle or shark. On average, for every 1,000 nets set around dolphin herds, fishers inadvertently capture 500 dolphins, 52 billfish, 10 sea turtles and no sharks. In contrast, typical bycatch from the same number of sets around floating objects includes only two dolphins but also 654 billfish, 102 sea turtles and 13,958 sharks. In addition, many juvenile tuna are caught under floating objects.

One solution to the bycatch from nets would be to fish for tuna with poles and lines, as was practiced commercially in the 1950s. That switch would entail hiring back bigger crews, such as those laid off when the tuna fishery first mechanized its operations.

The recent reductions in the bycatch of dolphins and turtles provide a reminder that although the state of the world's fisheries is precarious, there are also reasons for optimism. Scientific grasp of the problems is still developing, yet sufficient knowledge has been amassed to understand how the difficulties can be rectified. Clearly, one of the most important steps that could be taken to prevent overfishing and excessive bycatch is to remove the subsidies for fisheries that would otherwise be financially incapable of existing off the oceans' wildlife—but are now quite capable of depleting it.

Where fishes have been protected, they have rebounded, along with the social and economic activities they support. The resurgence of striped bass along the eastern coast of the U.S. is probably the best example in the world of a species that was allowed to recoup through tough management and an intelligent rebuilding plan.

Recent progress provides added hope. The 1995 U.N. agreement on high-seas fishing and the 1996 Sustainable Fisheries Act in the U.S., along with regional and local marine conservation efforts, could—if faithfully implemented—help to guide the world toward a sane and vital future for life in the oceans.

FURTHER READING

BLUEFIN TUNA IN THE WEST ATLANTIC: NEGLIGENT MANAGEMENT AND THE MAKING OF AN ENDANGERED SPECIES. Carl Safina in *Conservation Biology,* Vol. 7, No. 2, pages 229–234; June 1993.

GLOBAL MARINE BIOLOGICAL DIVERSITY: A STRATEGY FOR BUILDING CONSERVATION INTO DECISION MAKING. Edited by E. Norse. Island Press, 1993.

UNDERSTANDING MARINE BIODIVERSITY. Report of the National Research Council's Committee on Biological Diversity in Marine Systems. National Academy Press, 1995.

SONG FOR THE BLUE OCEAN. Carl Safina, Henry Holt and Company, 1998.

Shrimp Aquaculture and the Environment

CLAUDE E. BOYD AND JASON W. CLAY

ORIGINALLY PUBLISHED IN JUNE 1998

Shrimp aquaculture, or farming, first became profitable during the 1970s and has since mushroomed into a widespread enterprise throughout the tropical world. Thailand, Indonesia, China, India and other Asian nations now host about 1.2 million hectares (three million acres) of shrimp ponds on their soil, and nearly 200,000 hectares of coastline in the Western Hemisphere have been similarly transformed. Though rare in the U.S., where fewer than 1,000 hectares are devoted to shrimp farming, at least 130,000 hectares of Ecuador are covered with shrimp ponds. The seafood produced in this fashion ends up almost exclusively on plates in the U.S., Europe or Japan.

Hailed as the "blue revolution" a quarter century ago, raising shrimp, like many other forms of aquaculture, appeared to offer a way to reduce the pressure that overfishing brought to bear on wild populations. Shrimp farming also promised to limit massive collateral damage that trawling for these creatures did to other marine species, 10 kilograms of marine life being caught routinely for each kilogram of shrimp taken from the sea. Unfortunately, neither of these benefits has, as of yet, fully materialized. And as the record of the past two decades of shrimp farming clearly shows, aquaculture often creates its own set of environmental problems.

DOWN ON THE FARM

Normally, shrimp mate in the ocean. A single female spawns 100,000 or more eggs at a time, and within 24 hours the eggs that are fertilized hatch into larvae, which soon start feeding on plankton. After the larval period ends (about 12 days later), the young shrimp migrate from the open ocean into nutrient-rich estuaries, where they grow into more robust juveniles. Later they return to the sea to mature and mate.

For the most part, shrimp farming attempts to duplicate this natural life cycle. Aquaculturists induce adult broodstock to spawn in hatcheries by manipulating lighting, temperature, salinity, hormonal cycles and nutrients. After the eggs hatch, the managers of the hatcheries quickly transfer the offspring to rearing tanks where they can mature. During the early stages of this process, the tiny shrimp feed on microscopic algae. After the larvae grow bigger, they receive brine shrimp and manufactured feed. The managers keep the young shrimp in rearing tanks for an additional three weeks or so before releasing them into larger ponds.

In southeast Asia, most shrimp ponds are stocked with such hatchery-produced young. But in Latin America, many shrimp farmers prefer to raise larvae caught in the wild, because they are thought to be stronger and survive better in ponds. So the price for wild progeny may be more than twice that of shrimp conceived in a hatchery, and armies of collectors take to the water with nets to capture young shrimp for sale to the farmers. It is not clear whether fishing out so many larvae has depleted populations of wild shrimp. Still, in Central America, some commercial shrimp trawlers report that their catches declined noticeably when people began collecting larvae in large numbers from nearby estuaries.

Although fishing for shrimp larvae provides much needed work for many locals, their fine-mesh nets harvest essentially everything in their path, and inadvertent taking, or "bycatch," becomes a serious problem. The statistics are difficult to verify, but some workers believe that for every young shrimp snared in the wild, 100 other marine creatures will be killed.

Other environmental problems can arise from the ponds themselves. These shallow bodies are usually built by constructing earthen embankments along their perimeter. They vary in size from a few hundred square meters to many hectares, with average depths that are typically less than two meters. Usually, shrimp farmers pump seawater into canals from where it can then flow by gravity into ponds located somewhat inland, although some small-scale operations rely on the tide for filling ponds perched close to the sea.

The location of shrimp ponds is perhaps the most critical factor in controlling their impact on the surrounding environment. In Ecuador, ponds were initially constructed on salt flats and some other areas well suited to this use. That is, they were situated in places that were not particularly important for the proper functioning of the local ecosystem or for maintaining biodiversity. Yet as these expendable lands became scarce, shrimp ponds began to invade what was, from an environmental standpoint, more valuable property—wetlands and coastal thickets of salt-tolerant mangrove trees. In Thailand and many parts of Asia, shrimp aquaculture was never limited to salt flats. Although larger operations tended to avoid mangroves, about 40 percent of the small-scale farms—facilities set up with little forethought or investment capital—displaced mangroves.

Mangroves and wetlands are extraordinarily important both for the environmental services they provide and for the many plant and animal species that depend on them. For instance, mangroves soak up excess nutrients that would otherwise pollute coastal waters, and they provide protective nurseries for young marine animals. So the estimate of the United Nations Food and Agriculture Organization that about half the world's original endowment of mangrove forest has already been lost is quite troubling.

In most countries, the destruction of mangroves is driven primarily by people seeking wood for building or for fuel. Some mangrove-lined shores, like other kinds of forests, succumb to the pressures of development, which are often greatest along the coast. Shrimp farming alone appears to be responsible for less than 10 percent of the global loss. Yet in some countries shrimp aquaculture has caused as much as 20 percent of the damage to mangroves, and in some local watersheds shrimp farming accounts for nearly all the destruction.

INTENSIVE CARE UNITS

There are three primary methods for raising shrimp in ponds. These systems are classified according to the density of shrimp they contain, but they differ also in the nature of the feed used and in the rate of exchange of water between the ponds and the nearby ocean.

So-called extensive systems of aquaculture raise fewer than five shrimp for each square meter of pond water, whereas intensive systems grow 20 or more shrimp for each square meter of pond. Somewhere in between are the "semi-intensive" operations. The people who manage extensive systems of aquaculture nourish their charges by treating their ponds with fertilizers or manure to promote growth of algae. No other feed is given. In contrast, pellets made from plant and fish meals, nutritional supplements and a binder to enhance the stability of the feed in the water are applied daily to ponds undergoing semi-intensive and intensive management. Production during a 100- to 120-day crop is less than 1,000 kilograms per hectare (892 pounds per acre) in extensive ponds. Semi-intensive methods might produce as much as 2,000 kilograms per hectare, and intensive cultures can, in some cases, provide a phenomenal 8,000 or more kilograms per hectare.

On average, nearly two kilograms of food are needed to produce a kilogram of shrimp. Part of the reason for the inequality is that shrimp, like other animals, are not 100 percent efficient in converting food to flesh. Also, even in the best regulated feeding systems, up to 30 percent of the feed is never consumed. Consequently, a considerable amount of waste accumulates in the ponds in the form of uneaten feed, feces, ammonia, phosphorus and carbon dioxide. Usually, no more than a quarter of the organic carbon and other nutrients provided in the feed is recovered in the fattened shrimp at harvest. The excess nutrients stimulate the growth of phytoplankton, which eventually die, sink and decompose on the bottom of the ponds, consuming large amounts of oxygen in the process.

In traditional systems of aquaculture, the operators periodically remove the unwanted nutrients, dissolved gases, phytoplankton and pathogens by flushing them out to sea. In past decades, from 10 to as much as 30 percent of the water in the ponds was disgorged into the ocean each day. Today most shrimp farmers do better, exchanging daily from 2 to 5 percent of the pond water with the sea. Some shrimp farmers are attempting to eliminate this exchange completely. They have reduced the amount of wasted feed and also kept diseases in check by taking care not to stock their ponds too densely or with any infected larvae. In intensively operated ponds, mechanical aerators inject supplemental oxygen to prevent hypoxia from harming the shrimp.

The main chemicals put into ponds are fertilizer (to stimulate the growth of plankton on which the shrimp can feed), agricultural limestone and burnt lime (for adjusting the acidity of the water and underlying soil). In Asia, shrimp farmers also routinely add porous minerals called zeolites to remove ammonia, and they sometimes dose their ponds with calcium hypochlorite, formalin and some other compounds to kill pathogens and pests.

In some areas, the pollutants released from shrimp farms have exceeded the assimilative capacities of nearby coastal waters. Even if the quality of effluents from individual ponds falls within reasonable standards, too many farms in one area will eventually overwhelm natural ecosystems nearby, frequently causing unwanted fertilization (eutrophication) of coastal waters. The problem immediately spills over to all the coastal inhabitants—including the aquaculturists themselves, who must then struggle with the contamination of their own water supply.

But eutrophication is not the only threat. Viral diseases also haunt locales where concentrated shrimp aquaculture has degraded coastal waters. These diseases have sparked the collapse of much of the shrimp

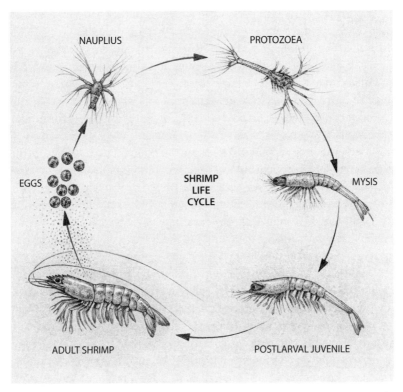

Shrimp life cycle, in which the eggs hatch and the shrimp grow through several stages before becoming adults, can be entirely duplicated by farmers.

farming in China and Taiwan, and they have caused serious difficulties in Thailand, India and Ecuador. The pathogens at fault can travel from country to country—even from hemisphere to hemisphere—in shipments of infected hatchery-produced shrimp. Diseases of shrimp can also be spread through uncooked, processed frozen shrimp.

Shrimp farmers have learned to fight these diseases in several ways. For example, they can now test the larvae they buy from hatcheries for dangerous viruses. And they have figured out how to dispose of shrimp from infected ponds so as to contain outbreaks. Certain hatcheries are using carefully bred broodstock to ensure that the larvae they produce are disease-free. Such advances are welcome, but some shrimp farmers also turn to using medicated feeds, a tactic for combating disease that may foster the proliferation of antibiotic-resistant bacteria or otherwise upset the local microbial ecology in worrisome ways.

RUNNING A TIGHTER VESSEL

Shrimp aquaculturists have recently started to address environmental concerns. Many of the rules of environmental etiquette are obvious. For example, ponds should not be constructed in sandy soil (unless impermeable clay or plastic liners are used) to prevent seepage of saltwater into freshwater aquifers. Discharge of effluents into predominately freshwater bodies or onto dry land should also be prohibited.

Making the proper choice of sites for the ponds is perhaps one of the easiest ways for shrimp farmers to limit environmental damage—at least for ponds that have not yet been built. There is no defense for putting shrimp ponds in mangrove forests or even in tidal wetlands. These areas are not suited for sustained shrimp farming: they often have soils that are incompatible with long-term shrimp production and, more troubling, are vulnerable to coastal storms.

Most large-scale aquaculturists have learned to do better, for themselves and for the environment, than to displace mangroves with their facilities. Instead they construct canals or pipelines to bring ocean water through the coastal mangroves to sites farther inland. And many smaller-scale shrimp farmers are forming cooperatives to pool the resources and knowledge needed for responsible operations. In Indonesia, some large producers are required by law to help small-scale shrimp farmers manage their ponds. It is imperative that such efforts expand so that shrimp farming neither causes nor takes advantage of the epidemic loss of mangroves.

Even shrimp farmers preoccupied with profitability should be able to understand the benefits of adopting better practices. It costs anywhere from $10,000 to $50,000 per hectare to build proper shrimp ponds. Abandoning these works after only a few years because they have been located inappropriately not only causes considerable environmental damage, it also proves needlessly expensive. So shrimp farmers would do well to pick suitable locations away from mangroves.

And there are other simple changes that would help both the environment and the bottom line. For example, farm managers commonly broadcast large amounts of food over their ponds once or twice a day; however, many smaller feedings at more frequent intervals, combined with the use of feeding trays, would require less food and cause less waste. Improved feeds—formulations that use greater amounts of vegetable protein and less fish meal—are more digestible, appear to last longer in the water and also produce less waste. Investing in these practices would discourage overfishing of the seas for shrimp food, and it would save shrimp

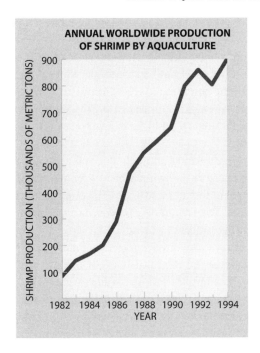

ANNUAL WORLDWIDE PRODUCTION OF SHRIMP BY AQUACULTURE

Shrimp aquaculture has been practiced for centuries, but only in the past two decades have people raised shrimp in massive quantities (*left*). The top shrimp-producing nations (*below*) straddle the equator in Asia and the Americas.

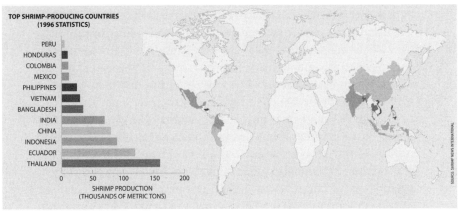

farmers money on feed, limit pollution and diminish the cost of cleaning up problems later. So it would also boost profits.

Another way to reduce water pollution is to avoid stocking ponds with too many juveniles and to restrict the amount of water exchanged with the sea. When the density of shrimp is right, natural processes within the ponds will assimilate much of the waste into the underlying soil. And although current technology requires ponds to be drained for harvest, operators could easily pass the water through settlement ponds to encourage denitrification and to remove many other pollutants associated with the suspended solids. Shrimp farmers should also refrain from mixing freshwater with the seawater in the ponds to reduce salinity. This practice (which, thankfully, has been abandoned by almost all shrimp farmers) is unnecessary and should be prohibited to avoid excessive drain on freshwater supplies.

Addressing threats to biodiversity is more difficult. But many problems could be eliminated if farmers raised only shrimp procured from hatcheries, carefully regulated the importation of broodstock and young shrimp, and cultured only native species. They could also prevent larger aquatic animals from being caught in their pumps by using intake screens.

Shrimp farmers should also pay more attention to the chemical additives they employ. Although most of the chemicals used in shrimp farming have a history of safe use, the application of chemicals other than agricultural limestone, burnt lime and fertilizers is usually unnecessary. In those rare instances where antibiotics are required, government regulators should evaluate the chemicals employed and prohibit potentially harmful ones—or at least make sure that they are used in a safe manner.

These governments should also require that careful studies of environmental and social impact precede the construction of new shrimp farms. That way, the communities involved could gauge the likelihood of damage to the local environment and identify conflicts in the use of land and water. Governments must also find ways to ensure that these initial efforts to protect the environment remain effective over time.

Although environmental impact studies would be valuable for new projects, many existing shrimp farmers will clearly need to change their practices. Here both a carrot and a stick are necessary. Some shrimp producers will see the wisdom of adopting more sustainable approaches themselves. In some instances, however, governments must impose regulations.

In all, shrimp farmers should welcome the changes on the horizon. Technological innovations promise to aid them in reducing the discharge of wastewater and extending the life of their ponds. Better breeding programs should offer varieties of shrimp with greater resistance to disease. The adoption of better practices will cost producers somewhat more in the short term. But in the long run these changes will pay for themselves by improving the efficiency and durability of their operations.

The shrimp industry seems to be responding to criticisms from environmentalists, and we are hopeful that shrimp aquaculture will prove much less harmful to the environment in the future. In fact, many of today's operations are better than those of the recent past in this regard. Yet the shrimp industry as a whole still has to evolve substantially before it attains standards that might allow shrimp aquaculture to flourish on the same site indefinitely. Only at that point will shrimp aquaculture join most other kinds of farming in achieving widespread acceptance.

FURTHER READING

SHRIMP FARMING AND THE ENVIRONMENT: A WHITE PAPER. C. E. Boyd. Shrimp Council, National Fisheries Institute, Arlington, Va., 1996.

WORLD SHRIMP FARMING 1996. Robert Rosenberry. Shrimp News International, San Diego, 1996.

MARKET POTENTIALS FOR REDRESSING THE ENVIRONMENTAL IMPACT OF WILD-CAPTURED AND POND-PRODUCED SHRIMP. Jason W. Clay. World Wildlife Fund, Washington, D.C., 1997.

TOWARD SUSTAINABLE SHRIMP AQUACULTURE. Jason W. Clay in *World Aquaculture,* Vol. 28, No. 3, pages 32–37; 1997.

The Evolution of Ocean Law

JON L. JACOBSON AND ALISON RIESER

ORIGINALLY PUBLISHED IN *SCIENTIFIC AMERICAN PRESENTS*, FALL 1998

Who owns the oceans? A quick glance at most maps of the world suggests that no one does. The oceans usually are depicted as an uninterrupted wash of pale blue, seemingly representing free seas subject to no nation's sovereignty or jurisdiction.

But this picture is inaccurate: it disregards the centuries-old political and legal struggles that have divided the oceans. To understand those tensions and the evolving international law of the sea, imagine instead an animated historical map of the world oceans in which one year whizzes by every four seconds.

In the early 17th century the waters appear calm, the seas a swath of undivided blue, while other colors battle for dominance over the continents. For the first 20 minutes (300 years), the oceanic parts of the map do not change much: only thin lines of fluttering color along the shores indicate national claims to the seas. When the map hits the mid-20th century, the waters suddenly explode in a fireworks display, starting at the coasts and expanding seaward. About two minutes later nearly 40 percent of the initial blue expanse is covered with many hues representing coastal nations. This final configuration reflects the oceans as they stand today.

The journey to establish this modern ocean geography has not been smooth sailing. Naval powers and coastal nations have fought to protect their watery domains and all the resources they contain—mineral and living, military and economic. Now, after decades of diplomacy, many nations have adopted an international law of the sea that outlines their rights and responsibilities regarding the use and management of the oceans. But whether agreement at the conference table means that cooperation will truly reign at sea remains an open question.

FREE SEAS

Scholars frequently mark the start of the era of free seas—the opening 20 minutes of running time on the imaginary animated map—with the

1609 publication of *Mare Liberum,* or "Free Seas." Written by Dutch jurist Hugo Grotius, the document supported the right of the Dutch East India Company to send its ships through seas claimed by Portugal. Grotius argued that the ocean was too wild to be occupied by nations and that its limitless resources made ownership absurd.

Although *Mare Liberum* itself did not necessarily spawn the age of free seas, its arguments were generally embraced by European nations that were pursuing profit and colonization throughout newly discovered regions. By the mid-17th century it was common practice for maritime nations to use the open ocean freely for the passage of vessels and for fishing.

The only widely recognized exception to the rule of free seas relates to the "territorial sea," a narrow offshore belt of national authority bordering the coast. By the 18th century the maximum breadth of this territorial sea, long subject to dispute, began to settle on a value of three nautical miles (around five kilometers). Although this measure has often been attributed to the distance that a land-based cannon could supposedly fire a ball, it is probably based on the length of the English league.

Within the territorial sea, each coastal country had nearly complete authority over the waters and seabed—including living and mineral resources—and the airspace above. Foreign vessels on the surface were allowed the right of innocent passage—that is, movement that does not threaten the peace and security of coastal nations. Beyond these territorial seas, surface vessels and submarines were free to navigate the "high seas." Ships from all countries were also allowed to fish these blue waters. By the early 20th century even the newfangled flying machines were accorded the right to fly over this vast, unbounded area.

THE 200-MILE CLUB

The era of the free seas survived three centuries and two world wars. Although the U.S. emerged from World War II as a global naval power with a consequent interest in preserving the broadest range of liberty on the seas, just weeks after the end of the war, America triggered a revolution in international law that led to the serious erosion of freedom of the seas.

In September 1945 President Harry Truman issued two proclamations pertaining to the oceans off U.S. coasts. One addressed the management of national fisheries beyond the territorial sea and the other, often referred to as *the* Truman Proclamation, claimed exclusive U.S. jurisdiction and control over the natural resources of the continental shelves adjacent to U.S. coasts—areas that, in many places, extended far beyond the outer

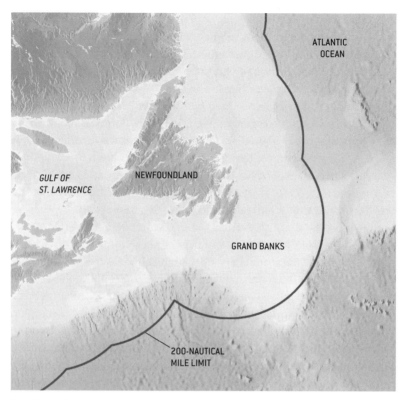

ATLANTIC
OCEAN

GULF OF
ST. LAWRENCE

NEWFOUNDLAND

GRAND BANKS

200-NAUTICAL
MILE LIMIT

Grand Banks off Newfoundland, Canada, have historically provided rich fishing grounds. Although these fisheries are now largely depleted, the parts of these shallows that lie just beyond the limit of Canada's exclusive fishing zone once attracted distant-water fishing vessels from many countries.

limit of the three-mile territorial sea. This assertion of national authority was illegal, but it was quickly approved and emulated by the international community. The proclamation thus created a doctrine of customary international law that recognized the exclusive right of coastal nations to control and extract the natural resources of the continental shelves off their shores.

Although it championed the traditional freedoms of navigation, the Truman Proclamation initiated a reaction that eventually turned the old Grotian order into a shambles. Many factors fueled this historic overthrow. Perhaps the most important was the suspicion and resentment raised by technologically rich nations, including the Soviet Union and Japan, that used their distant-water fishing fleets to harvest fish from the high seas just offshore from many foreign lands. To protect their coastal fisheries, several countries extended their territorial seas to 12 nautical miles.

Others carried this trend of expansion much farther. In 1947 Chile and Peru each asserted national control over the resources of the ocean and the seabed out to 200 nautical miles from their coasts. And in 1952 Ecuador joined its southern neighbors and claimed its own 200-mile-wide zone. Despite protest from various seafaring nations and those with distant-water fishing fleets, the so-called 200-mile club gradually added to its membership, first from Latin America and then from Africa.

In the midst of growing confusion over the state of international ocean law, the United Nations convened the first two conferences on the "Law of the Sea" in Geneva. The first one, conducted in 1958, adopted four conventions designed to codify and establish a set of principles and rules for sharing the oceans of the world.

The package of conventions painted a rather traditional view of marine geography, with the waters divided into high seas, territorial seas and a contiguous zone—extending beyond the territorial sea for no more than 12 miles—within which coastal countries could exercise limited jurisdiction for customs control and other defensive actions. But the countries recognized a new continental shelf doctrine.

The delegations were unable, however, to agree on a rule establishing the maximum outer limit of the territorial sea. (A second conference, held in 1960, also failed to reach a consensus on this issue.) Certainly none of the 1958 Geneva conventions on ocean law endorsed anything as extreme as the 200-mile claims asserted by a growing number of coastal nations. Instead countries attending the conference agreed to promote cooperation between coastal countries and distant-water fishing nations in the conservation and management of fisheries. Unfortunately, none of the most important distant-water fishing nations chose to join this particular convention, preferring instead to enjoy the customary freedom to fish the high seas.

THE SEABED QUESTION

Despite the Geneva accords, national expansion into the seas continued apace. By the mid-1960s this trend became so alarming to the two main naval powers of the day, the U.S. and the Soviet Union, that these cold war adversaries began to plot together to hold back the expansionist tide. The U.S. and Soviet Union feared creeping jurisdiction—the prospect that nations making claims over fish and other natural resources might also try to interfere with navigation and overflight in broad areas. The U.S. was especially concerned that the extension of territorial seas would inhibit

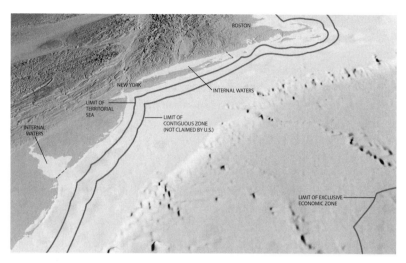

Offshore waters are divided by the 1982 United Nations Convention on the Law of the Sea into a territorial sea (which stretches 12 nautical miles from the coast), a contiguous zone (out to 24 miles) and an exclusive economic zone (to 200 miles). The convention also equated the minimum legal boundary of the continental shelf with the limit of the exclusive economic zone. Yet the physical continental shelf rarely extends that far, as can be seen in this map of the seafloor off the northeast coast of the U.S.

the passage of its ballistic-missile nuclear submarines—the main cold war deterrent to nuclear exchange with the U.S.S.R.—through such vital choke points as the straits of Gibraltar and Hormuz. The establishment of territorial seas even 12 miles wide would cause these and other straits to be blanketed by coastal waters in which, under traditional rules, submarines would be required to surface.

While the superpowers plotted, Ambassador Arvid Pardo of Malta addressed the delegates at the 1967 annual meeting of the U.N. General Assembly. He reminded the gathered members that recent investigations had shown that vast areas of the deep seabed—most beyond national jurisdiction—were literally paved with nodules containing valuable minerals, such as nickel, copper and manganese. Pardo urged the assembly to declare the deep ocean floor the "common heritage of mankind" and to see that its mineral wealth was distributed preferentially to the poorer countries of the global community. The General Assembly responded by adopting resolutions embodying Pardo's noble vision and calling for a new U.N. conference on the Law of the Sea.

The diplomats were charged to develop the concept of common heritage and to create a scheme for mining the seabed. But by the time the conference convened in New York City for its first session in 1973, its agenda had expanded to include nearly every conceivable use of the ocean, including fishing, navigation, protection of the marine environment and freedom of scientific research.

THE NEW MAP

The Third U.N. Conference on the Law of the Sea (dubbed UNCLOS III) was the most ambitious lawmaking endeavor ever undertaken by the international community. In a series of sessions that spanned nearly a decade, diplomats juggled and balanced the multitude of interweaving and highly politicized maritime concerns of more than 150 nations. The major naval powers and those countries with distant-water fishing fleets vied with coastal nations, and potential seabed miners argued with developing countries over control of the seafloor. In the end, UNCLOS III generated a complex constitution that regulated all human activities on, over and under the 70 percent of the planet that is covered by seawater.

The resulting treaty—the 1982 U.N. Convention on the Law of the Sea— endorsed the authority of coastal nations to govern an array of maritime activities within an area up to 200 nautical miles from their shores. In their territorial seas, which were to extend no farther than 12 nautical miles from the coast, these nations would retain their traditional sovereignty over all activities and resources but allow the right of innocent passage for foreign ships. In addition, the convention established exclusive economic zones (EEZs) that extend beyond the territorial sea to the 200-mile limit. Within their EEZs, coastal nations would now have exclusive control over the management of fisheries and other resources, subject to international duties of conservation and of sharing "surplus" fish. These nations would also have extensive rights and jurisdiction concerning such activities as marine scientific research and the construction and operation of artificial islands.

The contiguous zone, first recognized in the 1958 Geneva conventions, was extended to 24 nautical miles from shore. Within this zone, which overlaps the EEZ for 12 miles beyond the territorial sea, a coastal nation is allowed to enforce its laws on customs, immigration, sanitation and fiscal matters.

The 1982 treaty also expanded the boundaries of the legal continental shelf. Coastal nations would now have the right to exploit the natural resources of the seabed and subsoil as far out as 200 miles—or even beyond, to the edge of the entire shelf, slope and rise of the physical conti-

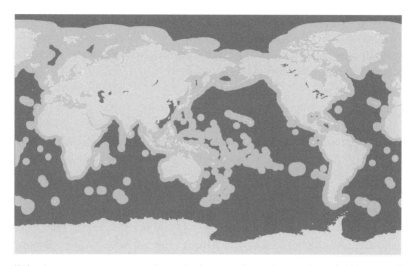

National governance zones surround every landmass on the earth except Antarctica. Most coastal nations claim the waters and seabed that lie within this area as their exclusive economic zone, within which they maintain exclusive control over the management of fisheries and other resources. Some nations claim these waters as fishing zones or as territorial seas.

nental margin. This right holds even when that margin extends beyond 200 nautical miles, as it does in several parts of the world.

The establishment of EEZs and the expanded definition of "continental shelf" constituted a major victory for the developing countries that formed the 200-mile club after World War II. The final impetus for global acceptance of the 200-mile zone, however, came not from UNCLOS III but from the U.S. Congress. In 1976 it established a 200-mile exclusive fishing zone for the U.S., causing a cascade of similar claims around the world. By the time the convention was adopted in 1982, the 200-mile concept had become customary international law.

The interests of the naval and maritime powers did not get subsumed in this process of national enclosure of the seas. These countries successfully negotiated for freedom of navigation and overflight within all EEZs. Moreover, the 1982 convention established a set of rules that would permit submarines to pass submerged through narrow straits, even those in which the waters consist only of territorial seas. Although the convention authorized island nations to designate the sea spaces within their island groups as "archipelagic waters," it granted foreign vessels and aircraft the freedom to navigate through them.

The convention also included many complex provisions on the marine environment. It expanded the rights of port nations and other coastal countries to guard against an influx of contaminants, and it declared that all countries would be responsible for protecting the marine environment from pollution, including that originating from sources on land.

In the end the 1982 U.N. Convention on the Law of the Sea was adopted in the General Assembly by a vote of 130 to 4 (with 17 abstentions), a stunning and unprecedented achievement for the community of nations.

U.S. BALKS

Although the delegations had agreed on the treaty as a package deal, the U.S. objected to the provisions on mining the deep seabed. The delegates' attempt to actualize Ambassador Pardo's grand vision for sharing the common heritage of mankind had created a legal quagmire so controversial and massively complex that in 1982 the U.S. rejected the entire treaty.

The U.S. and other mining nations would have undoubtedly preferred that the convention establish a simple registry system that would limit overlapping mining operations. Part of the proceeds from mining the areas of the seabed that lie beyond national jurisdiction could then be deposited in a special fund to be distributed to the poorest countries of the world. Instead the 1982 convention established the International Seabed Authority, headquartered in Jamaica, to which mining nations would apply for a lease. Miners would have to pay substantial up-front fees and royalties to the special fund and provide the technology and financing for the International Seabed Authority to mine an economically similar site in parallel. American free-marketeers strongly objected to this scheme.

Instead of signing the Convention on the Law of the Sea, President Ronald Reagan declared that those parts of the treaty concerning traditional uses of the ocean—including rights of navigation and fishing—were consistent with the customary law and practice of nations. And in 1983 he issued a presidential proclamation that established a 200-mile EEZ within which the U.S. would exercise all the rights and responsibilities recognized in the convention and other international customary law. The U.S. could thus arguably take advantage of the legal protections that the treaty afforded without endorsing the convention. (In 1988 the U.S. extended its territorial sea from three to 12 miles.)

Because the U.S. opposed the seabed-mining provisions, it appeared for a time that the convention might never be ratified by the 60 countries needed to turn it into law. But by 1994 the U.N. secretary general had worked out a separate agreement on deep-seabed mining, effectively replacing the mining provisions that had so offended the U.S. delegates.

President Bill Clinton subsequently sent the convention and the mining agreement to the Senate, where they currently remain awaiting approval. By that time, enough countries had ratified the convention to allow it to enter into force, at least for participating nations, in late 1994.

FISHERY STORMS

During the 1980s, while diplomatic attention was focused on the international control of deep-seabed minerals, a storm was gathering over the management of a much more significant marine resource: fisheries. Distant-water fishing fleets had responded to their exclusion from EEZs by perfecting techniques that would allow them to exploit the prized species that roam the high seas, where freedom to fish is still the rule.

Other fishing fleets opted to take advantage of places around the globe where valuable fish inhabited rich, productive waters just outside the 200-mile limit, such as parts of the Grand Banks off Newfoundland, Canada. They also homed in on coastal stocks of fish that swam through so-called doughnut holes, areas that are surrounded by but not part of EEZs.

Coastal nations whose stocks of fish were most vulnerable to these accidents of marine geography agreed that stronger international regulations were necessary. When discussions at the 1992 Earth Summit in Rio de Janeiro failed to resolve the issue, the U.N. General Assembly convened a new round of discussions to improve the management of high-seas fishing.

In March 1995, when the fish talks seemed stalled, a Canadian patrol boat arrested a Spanish trawler on the high seas for exceeding the internationally established quotas for the Grand Banks. This bold act signaled that at least one prominent coastal nation was willing to take the law into its own hands to protect the fish, even though the violation occurred outside the Canadian 200-mile limit and the forceful response threatened to undermine years of maritime diplomacy. Within a few months, the U.N. had adopted a new agreement on international fisheries to strengthen the standards by which nations collectively manage fishing on the high seas. That agreement, not yet in effect, calls for more careful setting of quotas for fish landings. Further, it allows coastal nations to inspect any vessel fishing on the high seas to ensure that it adheres to international regulations.

The effectiveness of this new agreement, and of the Law of the Sea in general, will depend on the willingness of many nations to be bound to its principles. We hope their commitment to international cooperation proves strong enough, even with few penalties and no high-seas police to enforce the rules.

FURTHER READING

ORDERING THE OCEANS: THE MAKING OF THE LAW OF THE SEA. Clyde Sanger. University of Toronto Press, 1986.

THE NEW LAW OF THE SEA. Edited by Christopher C. Joyner. Special double issue of *Ocean Development & International Law,* Vol. 27, pages 1–179; January–June 1996.

THE INTERNATIONAL LAW OF THE SEA. E. D. Brown. Dartmouth Publishing, 1994.

INTERNATIONAL FISHERIES LAW. OVERFISHING AND MARINE BIODIVERSITY. Alison Rieser in *Georgetown International Environmental Law Review,* Vol. 9, No. 2, pages 251–279; Winter 1997.

COASTAL OCEAN LAW. Third edition. Joseph Kalo, Richard Hildreth, Alison Rieser, Donna Christie and Jon L. Jacobson. West Publishing, 1998.

Fishy Business

SARAH SIMPSON

ORIGINALLY PUBLISHED IN JULY 2001

Cyanide is one of the fastest-acting poisons known to science. Once ingested, it cripples the body's ability to transport oxygen and begins asphyxiating tissues almost instantly. At higher dosages it slows the heart and even stops electrical activity in the brain. Given cyanide's lethal nature, it is difficult to imagine that squirting the substance at coral-reef fish is a good way to catch them alive. And yet that's common practice in the Philippines and Indonesia, whose collectors supply some 85 percent of the tropical fish that enliven the world's saltwater aquariums.

Disabling agile fish with cyanide makes it easier for divers to capture them before they hide among branches or crevices in the coral, but the consequences are severe. Some experts estimate that half of the poisoned fish die on the reef, and 40 percent of those that survive the initial blast are dead before they reach an aquarium. This startling mortality rate doesn't encompass the devastation to the living corals, invertebrates and nontarget fish in the path of the toxic plume.

Cyanide fishing is only one of several human activities—including poor forestry practices and industrial pollution—that are destroying coral reefs worldwide. But to many marine biologists, cyanide is one of the biggest dangers in Southeast Asian waters. The region harbors nearly 30 percent of the planet's coral reefs and boasts the greatest diversity of marine life anywhere—at least for now. According to two regional surveys published last year, only 4.3 percent of Philippine reefs and only 6.7 percent of those in Indonesia are still in excellent condition. And it is those reefs that live-fish collectors typically target.

For nearly 20 years, efforts to reform destructive aspects of the aquarium trade have fallen primarily on the shoulders of the export countries, with limited success as a result. Now a new strategy is placing more opportunity for reef preservation in the hands of importers, retailers and consumers along the trade route. In an ambitious campaign that could help save some of Southeast Asia's last pristine reefs, an international nonprofit organization called the Marine Aquarium Council (MAC) is developing a method for guaranteeing that the marine fish sold in pet stores are collected in an eco-friendly manner. By this fall MAC officials expect to have the first "certified" fish for sale in the U.S.

"There has never been a system to define, identify and verify environmentally sound practices and products in this industry," says the council's executive director, Paul Holthus. "We are also labeling these products so that the consumers can reward those who are responsible."

Because only a handful of the prized fish species can be raised in captivity, the fate of the aquarium hobby lies in preserving the reefs. Aquarium owners know this, Holthus explains, and that is why he believes they will demand certified fish—if given the choice. Even today retailers have no way of knowing the exact origins of the fish they buy from importers. For most of the history of the aquarium trade, people's choices have been limited by scant scientific evidence and by conflicting anecdotes about the severity and exact locations of cyanide use and other destructive activities.

TAINTED FROM THE START

Cyanide use in catching aquarium fish goes back nearly to the origins of the industry in 1957, when a Filipino entrepreneur shipped the first live fish to the U.S. in a tin can. Since the early 1960s, aquarium-fish collectors have squirted more than a million kilograms of cyanide onto Philippine reefs, according to estimates by the International Marinelife Alliance (IMA), a nonprofit organization founded in 1985 to fight the spread of destructive fishing practices in the region. Over the past 15 years, the organization has spent $1 million to train fishermen to use hand nets instead of poison. But progress is slow, explains IMA co-founder Vaughan R. Pratt, who directs the organization's operations throughout Southeast Asia. Adequate training can take several months, and until collectors become skilled with hand nets, they can earn more money using cyanide.

When news of cyanide fishing broke in the U.S. in the late 1980s, the gossip among hobbyists was that cyanide was a harmless anesthetic if used in the proper doses, Pratt says. Mortality rates of collected fish were often high, but for the most part aquarium hobbyists chalked that up to the notoriously fragile nature of the fish. Any number of problems along the trade route—poor water quality or too much time enclosed in a plastic bag, for example—can kill fish in transit.

Meanwhile the marine aquarium hobby was flourishing in the U.S. and Europe. Innovations in aquarium technology and animal husbandry

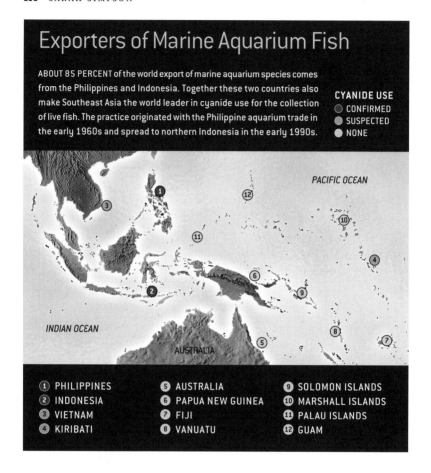

Exporters of Marine Aquarium Fish

ABOUT 85 PERCENT of the world export of marine aquarium species comes from the Philippines and Indonesia. Together these two countries also make Southeast Asia the world leader in cyanide use for the collection of live fish. The practice originated with the Philippine aquarium trade in the early 1960s and spread to northern Indonesia in the early 1990s.

CYANIDE USE
○ CONFIRMED
◐ SUSPECTED
● NONE

PACIFIC OCEAN

INDIAN OCEAN

AUSTRALIA

① PHILIPPINES   ⑤ AUSTRALIA   ⑨ SOLOMON ISLANDS
② INDONESIA   ⑥ PAPUA NEW GUINEA   ⑩ MARSHALL ISLANDS
③ VIETNAM   ⑦ FIJI   ⑪ PALAU ISLANDS
④ KIRIBATI   ⑧ VANUATU   ⑫ GUAM

were improving people's ability to maintain diversified tanks. This success boosted demand not only for fish but also for corals, anemones and other live reef species. According to a 1999 report by the South Pacific Forum Secretariat, an estimated 700,000 American households were keeping marine aquariums by 1992—a 60 percent rise in two years.

In the face of this increased demand for live fish—and because many Philippine reefs had been destroyed—cyanide fishing had spread to the northernmost island of Indonesia by the early 1990s. The most recent observations of IMA workers implicate nearby Vietnam and Kiribati as well.

For decades, reef-conservation workers on the front lines in the Philippines did not have the cooperation of the import countries to back their efforts. That is exactly what the Marine Aquarium Council has to offer. This past spring a 60-member MAC committee made up of representatives from industry, conservation, government agencies and academia outlined standards for managing the fish and the reefs in a sustainable way. The idea is to forge a reliable chain of custody in which fish are handled appropriately at each step of the trade route, from reef to retailer. One team is spending the summer motivating a string of collectors and exporters in the Philippines to comply with MAC standards, and another group is soliciting support from importers and retailers in the U.S.

FORGING AN UNBROKEN CHAIN

A promising point of origin for a certifiable trade route is the city of Bagac, about 90 miles west of Manila. This community of some 21,000 residents lies nestled between the South China Sea and the checkerboard pattern of bright green rice paddies along the flanks of Mount Bataan.

Of the city's 2,500 fishermen, perhaps 30 are aquarium-fish collectors who live with their families in a beachfront cluster of thatch and wood buildings. This group of men (only men fish here) has been collecting fish without cyanide for the past seven years. Before that, cyanide fishing was all they knew. The turning point for them was meeting IMA's Philippines field director Ferdinand Cruz.

Cruz, who is also a member of MAC, knows the aquarium trade in the Philippines as well as anyone. He was drawn to the fishing communities shortly after he and his sister and mother opened an aquarium-fish export business in Manila in 1984. Almost immediately the family was perplexed by the high death rate of the fish. "We thought our facility was at fault at first," Cruz recalls. When he visited his collectors, they hid their cyanide because it was illegal. Those who admitted to using the poison reasoned that the practice was a harmless way to catch fish alive.

Cruz wasn't convinced. He went out in the boats with the cyanide users and saw dead fish floating in buckets, dead fish on the seafloor and fish convulsing after being squirted. "Six months later I noticed that reefs that had been sprayed were dying and full of algae," he says. "I kept going back to the areas where cyanide was used and made my own opinion that it was a very damaging chemical."

Cruz worked for several years trying to keep his export warehouse cyanide-free, but he finally deemed that goal impossible to achieve. In 1993 he decided to abandon the business and began to work full-time for IMA. Since then, he has helped train some 2,500 of the estimated 4,000 aquarium-fish collectors in the Philippines. Cruz teaches them to set up barrier nets in canyons or deep fissures between coral heads and then herd the fish toward the net. Like most collectors in Southeast Asia, those in the Philippines breathe underwater through long, flexible plastic hoses called hookahs, which typically deliver air from an old compressor on board the fisherman's canoe. The diver holds the hookah in his teeth and often uses the bubbles from his exhalations to flush fish out of crevices in the coral and into the waiting nets.

Trained net fishermen are critical to a sustainable aquarium trade, but exporters also play a key role in MAC's plan. Most of the exporters, who constitute the next step in the chain of custody, are based in Manila. There, in warehouses filled with tanks, new arrivals typically mix with fish collected elsewhere around the country, many of them with cyanide. To make certification work, export warehouses will be required to quarantine fish that come from certified collectors.

At the warehouses, some fish can also be tested for cyanide exposure. Thanks in part to the efforts of Pratt and other IMA workers, cyanide detection laboratories are already in place. In 1991 the Philippines Bureau of Fisheries and Aquatic Resources contracted IMA to begin testing random samples of confiscated fish. The first detection laboratory opened near the Manila airport in 1991, and by early this year six laboratories around the country had tested more than 32,000 fish.

No current test can detect cyanide in living fish, so an unlucky few must be sacrificed. Chemists inspect and weigh each fish and liquefy it in a blender. The fish mush is distilled in a strong, hot acid so that any cyanide is liberated as hydrogen cyanide gas and then absorbed by a solution of sodium hydroxide. Electrode probes select for cyanide ions in the solution, enabling technicians to calculate cyanide levels in parts per million. Between 1996 and 1999, for example, workers saw the proportion of cyanide-tainted fish drop from 43 to 8 percent—a sign that IMA's investments are paying off.

Based on the considerable challenges of forging a certifiable chain of custody in the Philippines, Holthus says, the standards should be easy to maintain in Hawaii, Australia and other regions that already have high-quality operations in place. Once a chain of custody is certifiable in the export countries, it's up to importers and retailers in the U.S. to choose to buy those certified fish—and to live up to the MAC guidelines for their own handling practices. Even with the cooperation of importers, turning the poison tide in Indonesia, where fewer collectors are properly trained, will not be as easy. "If certification fails or only half-succeeds in the Philippines," Cruz cautions, "MAC standards will not take off in Indonesia."

Peter J. Rubec, who co-founded IMA with Pratt and works as a fisheries biologist with the Florida Marine Research Institute in St. Petersburg, hopes that the efforts of IMA and MAC will "provide the scientific evidence needed to convince the industry that net-caught fish are a viable economic alternative to cyanide-caught fish." Some retailers aren't so sure. "The pressure in the marketplace today is for lower prices, not higher ones," says James A. Bennett, owner of an aquarium retail store in Portland, Ore. "If MAC's plan increases the cost of the fish, that's not going to work."

Holthus hopes that certification will not actually cost the consumer any more. If the system works, he says, then the money saved by reducing fish mortality could offset any increased costs of certification. The MAC standards require that no more than 1 percent of each species die at any given point in the chain of custody. Achieving this goal, Rubec believes, would be a tremendous feat. He estimates that at least 10 percent of cyanide-caught fish die at each step in the trade route.

What is more, the fish grow considerably in value from one end of the trade route to the other. An orange-and-white-striped clownfish bought from a Filipino collector for about 10 cents, for instance, will sell for $25 or more in an American pet store. With that kind of markup, Rubec and others argue that the industry should be able to absorb the remaining additional costs of certification.

NOT SOON ENOUGH FOR SOME

Only time will tell whether economic obstacles will stymie MAC's mission. A few certified fish will be available to certain U.S. consumers this fall, but it may take a while for the desires of the market to force the aquarium trade to comply with the MAC standards. For some reef experts, the wait is agonizing.

"I don't think the Marine Aquarium Council has been tough enough," says marine biologist James M. Cervino, now a doctoral candidate at the University of South Carolina. After seeing cyanide damage for himself during his six years of service with the Global Coral Reef Alliance, Cervino argues that the trade should be halted temporarily: "If you don't have evidence that your fish were caught in a sustainable way, I can't see [this trade] being allowed to continue."

International law already bans the trade of thousands of species of stony coral under the Convention on the International Trade in Endangered Species of Wild Fauna and Flora (CITES), but most of the coral-reef animals in the aquarium trade are not listed. Some local village governments in the Philippines have experimented with export bans on certain live reef species, but Cruz says that the restrictions just drove the fishermen to other illegal activities. He has been campaigning for local governments to grant fishing licenses as an alternative way to regulate collection. "If this trade does not prove to be sustainable, then it will have to close completely," Cruz warns. "In the meantime, we should still use the resources the right way so that the community can profit from it."

After a certification system is up and running, import restrictions in the U.S. could tighten the loop. Last fall the U.S. Coral Reef Task Force, established by an executive order in 1998, helped to draft legislation that would ensure that consumer demand for marine aquarium organisms does not contribute to the degradation of reefs and their inhabitants, as it does today, says task-force member Barbara A. Best. The trade recommendations, which were still being considered by Congress in mid-May, reflect MAC's philosophy that certification is a way to encourage responsible and sustainable trade. The legislation also provides that after an unspecified period of time, the U.S. should ban the import of any coral-reef species unless it is accompanied by official documentation that the animal was not collected through the use of destructive fishing practices.

"Industry-certification schemes can be quite slow in catching on, and legislation that required certification would speed up the process," explains Best, who also advises the U.S. Agency for International Development on marine resource and policy issues. "I have had some retailers tell me that they view the trade recommendations as one way to ensure that everyone carries animals that are being collected sustainably and treated humanely," Best says. "This would also ensure that those retailers that are behaving responsibly and carrying certified products are not undermined by lower prices from other retailers."

"I would adapt, because all of my competitors would have to do the same," says Bennett, who has seriously considered eliminating sales of live marine fish from his Portland aquarium store. "Some of us would invest a lot of money in a hurry and try to farm these things."

Even with legislative restrictions and a strong consumer demand for certified fish working in tandem, coral reefs in certain export countries may still be at risk. Indeed, the first MAC-certified fish may not actually be cyanide-free. A few tainted fish may slip through this initial testing phase of MAC's long-term plan, in which the standards are intentionally basic so that they can be met relatively quickly. "We'll raise the bar as we go along," Holthus says. During the next two years, MAC will design more detailed standards, and the organization will monitor the health of the reefs as the changes take place.

Even if MAC's certification works to curtail cyanide use among aquarium-fish collectors, some researchers worry that there is still no guarantee that fish collection will not degrade the reefs. A case in point is Kona, Hawaii. Although aquarium-fish collectors do not use cyanide in this area, Brian N. Tissot of Washington State University in Vancouver, Wash., and Leon E. Hallacher of the University of Hawaii at Hilo discovered in late 1999 that the collectors' activities were stunting the populations of seven species of coral-reef fishes, three of which are herbivores. Without these grazing fish to keep algae in check, the prolific plants could eventually suffocate the coral animals.

Another challenge is reducing destructive practices among collectors of live food fish, who have spread cyanide use into Malaysia, the Marshall Islands, Papua New Guinea and possibly other areas of Southeast Asia.

Cruz and other IMA officials have reported that these fishermen often make forays into coastal areas where they have little interest in the long-term productivity of the reefs. Aquarium-fish collectors, on the other hand, are mainly people from local communities that have been relying on the same reefs for their livelihoods for generations. In part on Cruz's recommendation, MAC's certification standards require that local fishermen protect their own turf, even if that means patrolling coastal waters and chasing outsiders away—a practice that Cruz has already helped implement in several Filipino villages.

Creating strong incentives for local fishermen to be responsible for managing their own reefs "is probably the best hope in most of these areas for ever conserving the reefs," Holthus says. He has also seen growing interest among certain players in the live food fish trade to set up their own certification scheme. Better still, reefs might be developed into tourist areas for divers or protected parks where no fishing is allowed. But because of economic and political barriers, only a small number of reefs will ever fall into these categories.

The bottom line according to Cruz: If cyanide fishing isn't stopped, a lot of these reefs will be gone in a few decades. The good news, he believes, is that the battle against cyanide use in the Philippines is no longer uphill.

FURTHER READING

POISON AND PROFIT: CYANIDE FISHING IN THE INDO-PACIFIC. Charles Victor Barber and Vaughan Pratt in *Environment,* Vol. 40, No. 8, pages 5–34; October 1998. Also available at www.imamarinelife.org/environment.htm

CYANIDE-FREE, NET-CAUGHT FISH FOR THE MARINE AQUARIUM TRADE. Peter J. Rubec et al. in *Aquarium Sciences and Conservation.* Chapman and Hall (in press). Condensed version available at www.spc.org.nc/coastfish/news/lrf/7/LRF7-08.htm

Marine Aquarium Council, www.aquariumcouncil.org/

International Marinelife Alliance, www.imamarinelife.org/

U.S. Coral Reef Task Force, http://coralreef.gov/

Sharks Mean Business

R. CHARLES ANDERSON

ORIGINALLY PUBLISHED IN *SCIENTIFIC AMERICAN PRESENTS*, FALL 1998

I am on a wooden motorboat, over a small reef inside Ari Atoll, in the Maldive Islands. It is seven at night. Around me, the warm Indian Ocean is placid, and the lights of half a dozen small resort islands—Maayafushi, Halaveli and Fesdu, among others—glitter in the distance. In a few minutes I'll be in the dark water, scuba diving in the midst of feeding sharks.

I am neither as brave nor as foolhardy as that disclosure may suggest. The boat is moored in only eight meters (26 feet) of water, and I will be observing whitetip reef sharks (*Triaenodon obesus*). Inoffensive and fairly small, they are almost as far removed from "Jaws" as a shark can be.

A good shark dive is one of the great wildlife experiences, like a safari in East Africa or a cruise in Antarctic waters. And this particular spot in Ari's lagoon is one of the best shark-watching sites in all of the Maldives, a nation of some 1,200 tiny islands stretching about 1,000 kilometers southwest of the southern tip of India. Unfortunately, though, such places are becoming increasingly scarce.

A few days earlier I had visited Fish Head (Mushimasmingili Thila), another dive site in the same lagoon. Until recently, it was the premier shark dive site in the Maldives, being home to about 20 gray reef sharks (*Carcharhinus amblyrhynchos*). Stocky and mean-looking, they are nonetheless dedicated fish eaters, so they thrill divers without really endangering them. When I visited Fish Head last time, however, there was only a solitary shark in residence. Local fishermen had taken all the others. Strands of fishing line caught on the reef remained as evidence of their visits.

Operators of local dive shops are not happy about the loss. The warm, clear waters, extensive coral reefs and abundant sea life—including sharks—are the main attractions for the tens of thousands of divers who flock to the Maldives every year and form the backbone of the country's major industry: tourism. One estimate puts the annual number of dives made by visiting tourists at more than half a million; each dive costs

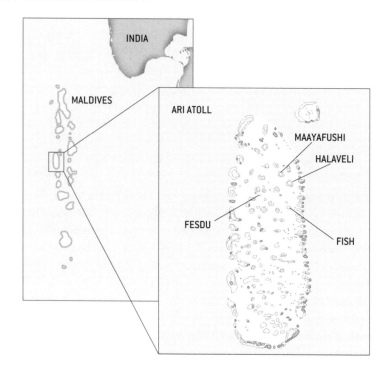

roughly $35. From the Maldivian perspective, that is a lot of money—the country's gross domestic product was only $423 million in 1995.

Apart from tourism, the only industry of any importance is fishing. Maldivian fishers have traditionally targeted tunas such as skipjack and yellowfin using the same live-bait pole-and-line method that their ancestors used 1,000 years ago. This preference for tunas, which are caught out at sea, left the reef-dwelling fish essentially undisturbed—until relatively recently. Over the past 15 to 20 years, East Asian buyers have encouraged Maldivian fishers to turn to the reefs. Sea cucumbers, groupers, giant clams and reef sharks such as the grays have all taken a hammering. As is true for countless other idyllic islands, two important industries—diving and fishing—are on a collision course in the Maldives.

DINNERTIME ON THE REEF

Such developments are the backdrop for my reef dive in Ari Atoll. After struggling into my scuba gear in the darkness, I duckwalk off the bow and swim down through the cloud of bubbles caused by my entry. Almost immediately my flashlight beam picks up the reef below.

The reef top is a seething mass of fish. Bluestreak fusilier fish (*Pterocaesio tile*) are milling over the reef and among the rocks, like a living carpet. These fish feed by day, forming great schools above the reef where they peck at incoming plankton. At night the fusiliers sleep in crevices in the reef. Now, just after dusk, they are trying to settle down for the night.

Out of the darkness a shark appears. Then another and another. They are whitetip reef sharks; at about a meter and a half long, they are hardly the efficient killers of popular imagination. Perhaps dazzled by my dive light, they bump into rocks and snap their jaws shut on empty water a full second after startled fusiliers have darted off. Eventually, and right in front of me, one whitetip bites down on a fusilier and with much head-shaking makes off with its dinner. The fresh wounds on many other fusiliers attest to recent and frequent near misses. After an hour watching a dozen or so sharks, I return to the mooring line and slowly ascend to my boat.

Although there's nothing like being next to feeding sharks to pump up the adrenaline, it took bloodless calculations to shed light on the conflict over reef resources. To estimate the value of sharks to the two industries, tourism and fishing, I did a survey of shark diving in the Maldives in 1992 with Hudha Ahmed, my colleague at the Ministry of Fisheries and Agriculture. We found that the money spent by divers on shark dives in the Maldives amounted to some $2.3 million a year. Some $670,000 came from dives at Fish Head alone.

We further estimated that for all shark-watching dive sites, the average value of a live gray reef shark was about $3,300 a year. Because the sharks can live for at least 18 years, and recognizable individuals have been seen at dive sites in the Maldives for many years in a row, the total value of each shark is actually several times higher. In contrast, a dead reef shark has a onetime value (cut up for meat and fins) of about $32 to a local fisherman. Thus, at dive sites, gray reef sharks are worth at least 100 times more alive than dead.

From an economic point of view, it clearly makes sense to ban shark fishing and leave the sharks as high-earning attractions for visiting divers. That, however, would deprive the fishermen—who gain few direct benefits from tourist income—of even the meager benefit conferred by dead sharks.

As one step toward protecting sharks and other marine life, the Maldivian government designated 15 popular dive sites as Marine Protected Areas in 1995. Eight of the sites, including Fish Head, were major shark-watching dive sites. Unfortunately, there was no means of enforcing the protected status of these areas. During 1995 and 1996, the shark population at Fish Head plummeted, as did the number of divers visiting the site. The loss of revenue is difficult to estimate, but my back-of-the-envelope calculations suggest that it is on the order of $500,000 a year. All this potential revenue was lost for the sake of about 20 sharks, which probably earned their captors less than $1,000.

What has occurred at Fish Head has happened at innumerable other locations around the world. The only difference is that I and other divers have monitored the change at Fish Head, and we are trying to do something about it. There is talk of banning shark-fin exports, but such a move would unfairly affect those fishermen who target oceanic sharks, which are more abundant than reef sharks. There is also talk of extending the system of protected areas, but lack of enforcement remains a problem.

The Maldives, moreover, are not the only trouble spot: Shark fisheries are in decline all over the world. Part of the problem is the sharks' biology: they grow slowly, mature late and have small numbers of young. Whitetip reef sharks, for example, mature at about five years and gray reef sharks at seven or eight. Both species give birth to typically two or three offspring at a time—a tiny brood in comparison to the hundreds or thousands of eggs produced by most bony fishes. As a result, shark populations are unusually sensitive to being overfished. Unfortunately, a quarter of the world's population craves shark's fin soup.

In many tropical countries, where shark fishery regulations are difficult to enforce, reef shark populations have dwindled to a small fraction of their original numbers. Even the lure of tourist dollars has not yet stemmed the losses. It is possible that we have already entered the twilight of one of nature's great and stirring spectacles: the sight of sharks feeding in the wild.

Fishing the "Zone" in Sri Lanka

ANTON NONIS

ORIGINALLY PUBLISHED IN *SCIENTIFIC AMERICAN PRESENTS*, FALL 1998

The sun has not yet risen on this Friday morning in March as I step on board the *Kamalitha,* a 10.4-meter (34-foot) fishing trawler tied to a dock in the harbor at Beruwala, on the Sri Lankan coast 55 kilometers south of the capital city of Colombo. Around the vessel, the busy little harbor bustles in the darkness as scores of other small but sturdy trawlers either are getting under way or are pulling up to unload their weary crews and the previous night's catch.

As we head out of the harbor, I duck into the boat's cabin to join my shipmates for the voyage. There is a tense moment as the skipper barks out crisp orders to the crew and we swerve to avoid a massive inbound ship. Then it is smooth sailing as we head southwest under a cloudless sky.

Dressed in a striped, cotton Henley shirt and madras shorts, the 29-year-old skipper, Kapila Nishantha, already has 17 years of experience fishing in Sri Lankan waters and has been a captain for the past five. His three-man crew—Gamini Silva, Palitha Dodampe and Vincent Vithana—is also seasoned, having been recruited into the same profession as their fathers and grandfathers. My role is that of an observer; as a newspaper reporter based in Colombo, I have written several articles on commercial fishing and am eager to see the industry from a new perspective.

As it turns out, I am not the only observer on board. A fisheries inspector, Susantha Wijesuriya of the Ministry of Fisheries and Aquatic Resources Development, has been charged with collecting data on the catch, which the ministry will use for statistical purposes. The data will also be put to scientific use: the ministry routinely shares information and works closely with the National Aquatic Resources Research and Development Agency (NARA), a governmental research center in Colombo.

The official interest reflects the fact that fishing has for many years been a prime foreign-exchange earner for Sri Lanka, raking in billions of rupees (tens of millions of dollars) annually. Some 100,000 Sri Lankans support their families comfortably as fishermen, and thousands more

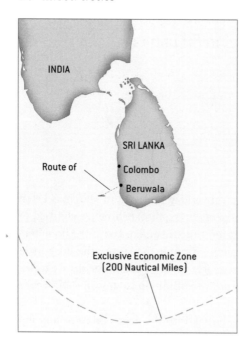

work in related jobs, such as mending nets and selling seafood to the resort hotels along the coast and in the central hills. Not surprisingly, the number of young people seeking to become fishermen is on the rise.

So far there has been plenty of fish for all. The total catch was around 217,000 metric tons in 1997. Tuna constituted almost half the catch, followed by shark at about 35 percent and billfish (marlin, sailfish and swordfish) at around 10 percent.

Ministry sources say the tuna varieties are actually increasing at present. A drop in the shark catch, however, which consists mostly of blue shark, has been observed over the past year. Officials hope that the blue shark fisheries will recover as larger boats and access to better bait allow fishermen to increase their catch of the more desirable tuna.

A more difficult problem involves poaching within Sri Lanka's 370-kilometer (200-nautical-mile) exclusive economic zone (EEZ), within which only Sri Lankan vessels have the right to fish. Estimates are that foreign trawlers, which are typically much larger and better equipped than their Sri Lankan counterparts, snatch up to 25,000 tons of fish every year from Sri Lanka's waters. For years, Sri Lankan officials had only two boats to patrol the country's 460,000-square-kilometer EEZ and could do little to stop the poaching. More recently, the administration of President

Chandrika Bandaranaike Kumaratunga has pledged to create a larger, better-equipped coast guard to address the problem.

Such concerns seem far away on the *Kamalitha,* as we push out to sea. With the sun rising into the sky behind us and the waves thudding on the hull, we are headed to a point about 90 kilometers from shore, well within the EEZ. Filling the time, Vincent rhapsodizes about a young woman in a neighboring village with whom he is smitten. Later, Gamini fires up the tiny stove in the cabin and begins preparing a meal of rice, vegetables and (what else?) fish curry. We eat heartily, the sea air having made us ravenous.

By three P.M., it is finally time to cast bait. The longlines go out first. There are five segments, tied end to end, for a total length of 2,500 meters. Evenly spaced buoys keep the lines near the surface. Along each 500-meter segment are some 25 branch lines, spaced at intervals of about 20 meters. Each of these vertical branch lines measures about 15 meters and has a baited hook at the end. Many of the hooks are baited with hunks of squid, a favorite of tuna. Scraps of beef are also used to lure sharks.

The longlines are followed by the gill nets, which are held in place between a top rope kept near the surface by polyurethane floats, and a weighted bottom rope. There are a total of 25 pieces of net, each measuring about 100 meters wide, suspended side by side between the top and bottom ropes. So when deployed, the net is like a wall of mesh 2,500 meters wide. A thick, 150-meter rope secures the end of the gill net to the stern.

With the lines and nets deployed, once again there is nothing to do but wait. To amuse themselves, the crew members sing or hop overboard for a swim. After the sun slips below the horizon, a chilly breeze blows past intermittently, reminding us that we are in water kilometers deep. Around us the murky ocean seems endless.

In the darkness on deck, the only light comes from the cabin's windows and, on the eastern horizon, the dim glow of coastal towns. Inside the cabin, crew members take turns napping on two narrow bunks. For skipper Nishantha, though, there is no rest. He spends most of the evening in the captain's seat, going out on deck from time to time to peer into the night.

In the early-morning darkness, the catch is hauled in. It amounts to almost 400 kilograms (about 880 pounds) and includes three tunas, each weighing around 25 kilograms, and some blue sharks. To keep them from biting, fish still struggling are clubbed on the head, blood spattering on the deck.

As the catch is tallied, Wijesuriya, the fisheries inspector, begins collecting data. For each fish he measures various dimensions, such as the length from snout to caudal fin, and notes the approximate location where it was caught and the type of fishing method that snared it. The type of net or length of longline (and number of hooks) are all recorded. According to Champa Amarasiri, the head of NARA's marine biology division, the data are used to estimate the age of the fish and to make inferences about the status of the fisheries.

Wijesuriya's presence on the *Kamalitha*, however, is unusual. With only 173 inspectors and a fishing fleet of 15,000 to 20,000 "day" and "multi-day" fishing boats in Sri Lanka, the fisheries ministry and NARA must rely on the fishermen themselves to log most of the data on their hauls.

For the captain and crew of the *Kamalitha*, data logging is already as much a part of their world as gill nets and compasses. And in the chilly, predawn darkness 90 kilometers out at sea, with the fish stored, the nets stowed and the data logged, it is time to return to port. Nearing the harbor, we join dozens of boats returning from all directions. As the fish are unloaded and sold or auctioned, other trawlers put out to sea, sleepy fishermen head home to their families and a new day dawns on Beruwala Harbor.

DANGEROUS WATERS

Giant Earthquakes of the Pacific Northwest

ROY D. HYNDMAN

ORIGINALLY PUBLISHED IN DECEMBER 1995

Few people question the possibility of a devastating earthquake once again hitting Los Angeles or San Francisco. The state of Alaska has also suffered some serious shaking, including, in 1964, one of the world's largest earthquakes. Until recently, however, many residents believed that the intervening territory from northernmost California to southern British Columbia (an area sometimes referred to as Cascadia) was a safer place to live. Seismologists had recognized that Vancouver and Seattle were not exactly sheltered—sizable earthquakes buffeted the region in 1946, 1949 and 1965—but no truly disastrous events had ever damaged these cities.

Yet views have changed drastically. Ten years ago Thomas H. Heaton of the U.S. Geological Survey and Garry C. Rogers of the Geological Survey of Canada began warning that giant earthquakes could indeed strike this seemingly quieter stretch of coast. Initially, many scientists questioned the seriousness of the threat, but most doubters now realize that such earthquakes have happened in the past and will do so again. How could perceptions have shifted so quickly?

To understand the change in thinking requires some knowledge of the way seismologists estimate how and where powerful but infrequent earthquakes occur. For most active fault zones, the rate at which earthquakes take place decreases with increasing size in a systematic way, as was shown in the 1930s by Beno Gutenberg and Charles F. Richter. This regular pattern applies up to some maximum earthquake size—one that corresponds to a break of the entire fault zone from end to end. Using the Gutenberg-Richter relation, seismologists can gauge how often large earthquakes strike a given place even if no such events have ever been recorded. Engineers can then design buildings, dams and other structures accordingly.

In a few areas, this strategy fails. Sizable earthquakes can hit without small ones, presenting seismologists with a vexing problem: How can the danger from large earthquakes be reasonably defined? This difficulty applies to Cascadia, where one of the tectonic plates underlying the Pacific Ocean thrusts underneath the coast of western North America in a process termed subduction. Although regional seismic activity can be quite intense in some areas inland of this coast, no earthquakes of any size have been detected where most of the motion is seemingly focused— on the main thrust fault that separates the Juan de Fuca plate from the North American continent.

In global perspective, the lack of such thrust earthquakes is surprising. Most subduction zones have experienced great thrust events (defined as those having a Richter magnitude higher than 8) at some time. These earthquakes are especially concentrated around the rim of the Pacific Ocean in a vast band called the ring of fire—a name that comes from the lines of active volcanoes that lie landward of where the oceanic crust dives into the earth's mantle.

NO LARGE EARTHQUAKES?

There are several possible explanations for the absence of major subduction earthquakes. Although the Cascadia part of western North America has many of the characteristics of a subduction zone, the Juan de Fuca plate may have stopped moving toward North America in geologically recent times. Twenty years ago, when geologists first debated this question, my colleague at the Pacific Geoscience Center Robin P. Riddihough and I wrote an article making the case that convergence and underthrusting are indeed continuing. The work of many researchers has since confirmed that the Juan de Fuca plate has not made a sudden stop. Persuasive evidence for continued motion comes from the study of sediments lying underwater at the base of the continental slope. These muds and sands were laid down in the deep sea as flat layers, but even the most recent deposits are found to be highly contorted. The North American continent, acting as a giant bulldozer blade, has scraped them off the oceanic crust and left them as crumpled evidence of continuing subduction.

Perhaps the most dramatic evidence for ongoing subduction came in 1980 with the volcanic eruption of Mount St. Helens in southwest Washington State. Scientists have recognized for many years that such volcanoes are a consequence of subduction. Some geologists had thought the Cascade volcanoes were dormant. But this volcanic cataclysm left little doubt that the Cascadia coast is indeed an active part of the ring of fire.

To reconcile the plates' convergence with the absence of thrust events, some scientists have supposed that the downward push below the coast involves a smooth, stable slide, not the jerky "stick-slip" behavior that generates earthquakes. The alternative explanation is that the fault between them is truly locked (the friction being large enough to hold the two plates firmly together), so that there is not enough movement to generate even small earthquakes. If the fault is freely sliding, the chance of large thrust earthquakes is slim. But if the fault is locked, the plate convergence must be accommodated by the silent but deadly buildup of strain in the rocks around the fault, the makings of a significant earthquake.

The lack of substantial earthquakes in the historical record might at first seem to favor the idea that the fault is slipping quietly. That interpretation, however, neglects the brevity of the record along this coast. Only a little more than 200 years ago did the explorers Juan Perez and James Cook first visit the region. The limited written history contrasts markedly with the span of Japanese records describing many large subduction earthquakes and the so-called harbor waves ("tsunamis") that usually resulted from them. That detailed archive extends back to the seventh century.

WHEN WAS THE LAST ONE?

To probe those times before Europeans arrived on these shores, researchers have sought traces of past earthquakes in the geologic record. They found some telltale evidence in sheltered inlets where salt marshes form between high and low tide. Excavations of these coastal marshes uncovered a remarkable record. Brian F. Atwater of the U.S. Geological Survey was first to show that distinct layers below the present marsh (spaced at successive depths of about a meter) contain peat that is made of the remains of vegetation identical to the flora now living in the intertidal zone. He concluded that each peat deposit constitutes a former marsh that was buried when the ground abruptly dropped with the release of strain in a sizable earthquake.

What makes his interpretation even more convincing is that many of the buried peat layers are covered by sand washed in by the huge tsunamis that rushed onto the subsided coast. Theoretical modeling as well as preserved geologic effects on the shoreline indicates that these waves attained heights of 10 meters on the open coast and much higher still in some confined inlets.

After the tsunamis dissipated, mud slowly filled the subsided region, and the marsh vegetation returned. Thus, the repeated sequences of peat, sand and mud clearly demonstrate that large earthquakes have plagued the region in the past. But how long ago were these prehistoric upheavals? The ages of the peat layers are difficult to determine precisely, but coastal fir trees have been found that were drowned by the ocean after the land abruptly subsided. By examining growth rings and measuring radiocarbon in these trees, researchers have estimated that they died in the last great earthquake, which hit the area about 300 years ago. Before that, similar events struck at irregular intervals of about 500 years.

This conclusion is also supported by unusual deposits found far out on the floor of the ocean. Scientists at the University of Oregon have sampled seafloor sediments in long core tubes and found fine-grained muds alternating with sandier layers. Mud is typical of the deep sea bottom; it accumulates from the slow, continuous rain of fine sediment settling from the ocean above. The sandier sediments, however, are strange to find far from shore. John Adams of the Geological Survey of Canada provided an explanation: energetic earthquakes could have triggered huge submarine landslides that carried coastal sediments down the continental slope and out onto the deep ocean floor.

The timing of the events is hard to judge from the sediments, but a peculiar deposit found near the base of some of the cores gives an important clue. This layer contains volcanic ash from the eruption of the former Mount Mazama in Oregon (now known as Crater Lake). That colossal explosion—similar to the recent Mount St. Helens blast—happened 7,700 years ago. Assuming that the rain of mud onto the seafloor was steady, the chronology for these earthquakes proves similar to the results from coastal peat deposits. The most recent event happened about 300 years ago, and the 12 previous submarine landslides were separated by 300 to 900 years.

A clever strategy may pinpoint the time of the most recent earthquake even more precisely. Tsunamis generated by Cascadia earthquakes with magnitudes near 9 should be large enough to be noticed in Japan even after traveling across the Pacific Ocean. Recognizing this fact, Kenji Satake and his colleagues at the Geological Survey of Japan think they have found the written record: a two-meter-high tsunami that washed onto the coast of Honshu nearly 300 years ago. After correcting for the time the wave would have taken to travel to Japan (and the time zone change), Satake determined that the earthquake occurred along the North American coast on January 26, 1700, at about 9 P.M.

Remarkably, that detective work agrees with reports of a disaster preserved in the oral history of the original residents of British Columbia. My colleague Rogers found what may be a description of this event in the provincial archives in Victoria. Native tradition records that an earthquake struck Pachena Bay on the west coast of Vancouver Island one winter night; in the morning the village at the head of the bay was gone. Gary A. Carver of Humboldt State University uncovered a similar account in the unwritten lore of northernmost California. Thus, native stories, Japanese writings and sedimentary deposits all point to the inevitable conclusion that giant earthquakes do in fact haunt the Cascadia coast.

THE GREAT EARTHQUAKE CYCLE

Like all earthquakes, large subduction zone events prove complex when considered in detail. The basic process, however, follows the simple "elastic rebound" theory first developed for the notorious San Andreas fault in California. According to this concept, ongoing movement between two plates compresses and bends the crust as stress accumulates. Contrary to the illusion of the earth being made of rigid and solid rock, the contraction is nearly elastic. If not squeezed too much, the earth acts like a gigantic piece of rubber. Eventually, however, the tectonic forces become so extreme that they exceed the hold of friction along the fault. The surface slips abruptly, and the elastic energy that was stored over many years radiates outward as ground-shaking earthquake waves. The fault then locks once more, and the cycle of tectonic stress buildup and release resumes.

Along the Cascadia subduction zone, the oceanic Juan de Fuca plate encroaches on North America by about 40 millimeters a year. This progress may seem slow, but it represents a considerable shortening—about 20 meters in a typical 500-year-long stretch between giant earthquakes. The motion is taken up by elastic shortening distributed across a swath several hundred kilometers wide. But the tectonic stresses cause more than just horizontal contraction—the ground moves vertically, too. As the oceanic plate dives under the coast, it drags the seaward nose of the continent downward and causes parts of the North American plate further inland to flex upward; this process mimics the bending of a long board over the edge of a table—as the front is forced down, a bulge forms behind. When a large earthquake breaks the locked fault, the seaward part of the continent springs back, and the bulge collapses. The abrupt rise of the outer continental shelf generates tsunamis, and the sudden fall of the "flexural bulge" centered near the coast causes the drop that buries intertidal salt marshes.

The position of the locked zone proves especially important because this surface becomes the source of seismic-wave energy when the fault eventually gives way in an earthquake. The landward limit of the earthquake source zone affects how closely the earthquake will impinge on the larger population centers; the seaward edge controls where tsunamis will

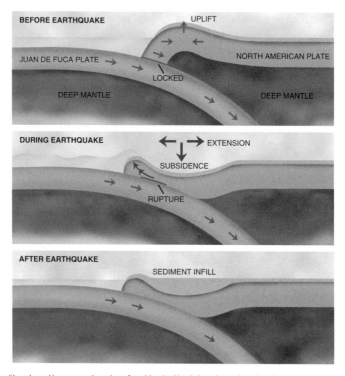

Flexing like a board bent over the edge of a table, the North American plate develops an inland bulge as its western margin is pulled downward by the oceanic slab (*top*). After an earthquake releases the stress, the bulge collapses (*middle*), forcing much of the coastal region to subside and fill with sediment (*bottom*).

develop. The total width of the source zone influences the seismic hazard because it sets the maximum size of the earthquake.

Scientists can determine the extent of the locked zone from the form of crustal deformation. If the locked zone is narrow, extending only a short distance down the inclined fault, the region of elastic bending will also be quite restricted. Conversely, if the locked zone runs appreciably farther down, the bending deformation will reach a long distance inland. Land surveying can thus help to map out the earthquake hazard. The rates of deformation are only a few millimeters a year, but they can be resolved with modern surveying techniques if the measurements are applied with exceptional care.

WATCHING THE STRAIN BUILD UP

Several different kinds of observations repeated over time define how the Cascadia margin is currently deforming. Geophysicists can follow the horizontal shortening of the coastal region by measuring, for example, the distance between surveyors' benchmarks on mountaintops using a laser ranging device. This feat requires a good deal of care and plenty of clear sky (not common in the rainy West Coast mountains). Using this technique, James C. Savage and his colleagues at the U.S. Geological Survey first reported in 1981 that the crust near Seattle was shortening perpendicular to the coast. They concluded that strain was building toward an appreciable earthquake.

Some survey methods are sufficiently sensitive to vertical motion. The most simple of these, known as leveling, employs the same technique one sees being used along highways. Surveyors take sightings on calibrated rods to measure the difference in elevation between two places. By combining the measured offsets, surveyors can determine relative heights throughout a network of connected points spread over large distances. Repeated surveys after several years yield the uplift or subsidence of one position with respect to another. The Geodetic Survey Canada has, for instance, carried out several surveys of exceptional accuracy specifically to study earthquake-related uplift. One of these field experiments tracked back and forth across the width of Vancouver Island (about 100 kilometers each way) in a series of sightings, each of 100 meters or so. The complete circuit had a total vertical error limited to only one centimeter.

Another method makes use of tide gauges that track the level of the sea relative to coastal bedrock. The primary purpose of these devices is to monitor the ocean, but surprisingly, with gauges that have been recording for 20 years or more, it is possible to use the average level of the sea surface as a reference and to trace subtle vertical shifts of the land. The record must, of course, be long enough to smooth over tides and other oceanographic variations, such as El Niño, that can endure for years. One also needs to account for the steady global rise in sea level (of about two millimeters per year) and to correct for postglacial rebound, the slow but continuing rise of crust initially pushed downward under the weight of glaciers from the last ice age.

Yet a third way to detect vertical motion is by using gravity, a force that varies with the square of the distance from the center of the earth. Although it is impossible for people to sense the slight shifts in their weight when they change altitude, sensitive instruments can register these variations. By repeatedly measuring gravity at one spot every few years, geophysicists have been able to estimate the rate of coastal uplift.

During the past few years, the satellite-based Global Positioning System (GPS) has permitted scientists to measure distances and vertical offset between sites spaced hundreds of kilometers apart. Herb Dragert and

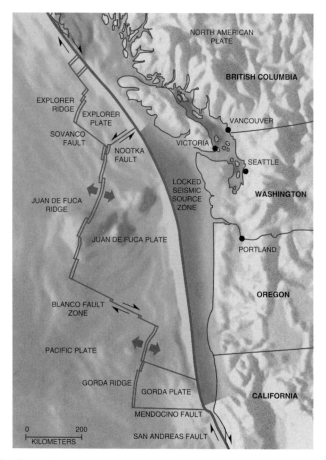

Converging plates are locked together over a confined region of the thrust fault under the coast. The extent of this locked zone is limited on its western side because clays deposited on the surface of the downgoing oceanic crust help to lubricate the fault. To the east, the locked zone gradually fades to the point where the deeply buried fault slides freely because of elevated temperatures.

Michael Schmidt of the Geological Survey of Canada have used GPS to show that every year coastal Victoria shifts nearly a centimeter closer to Penticton (a locale some 300 kilometers inland). GPS is accurate and inexpensive and in the future may prove the most effective technique for keeping track of the subtle bending and squeezing of the earth's crust that leads to earthquakes.

All these methods give similar results: the Cascadia margin currently rises by one to four millimeters a year, and it also contracts horizontally by several centimeters every year. This deformation—direct evidence that the crust is being squeezed between converging plates—registers the slow but relentless accumulation of strain that is building toward the next catastrophic release.

A TROUBLEMAKER LOCKED UP

Because knowing the position of the locked part of a fault is so critical to defining earthquake risk, my colleagues at the Pacific Geoscience Center and I have tried to determine the extent of the locked zone by comparing survey measurements with mathematical models of the deformation. Fitting the observations to theory allowed us to map the width of the locked zone deep below the earth's surface. The actual situation is somewhat more complex than this simple conceptualization: at the deeper, landward boundary of the locked zone, there is a gradual transition between areas that are rigidly locked and those that are completely free-sliding.

The comparison between our data and models shows that for most of the Cascadia coast the locked zone is restricted to a swath 50 to 100 kilometers across that runs underneath the continental shelf. (It widens considerably only near the coast of northern Washington.) This surface represents a huge fault area with potential for enormous earthquakes. Yet, curiously, it is unusually narrow compared with other subduction zones.

Such differences prompted us to examine which attributes of the geology influence the width of the locked zone. Many factors may contribute, but temperature plays a dominant role. For example, the clay-rich sediments that blanket the oceanic plate may act to lubricate the seaward, "up-dip" edge of the fault; as the ooze becomes more deeply buried, however, the clays chemically alter into stronger minerals that prevent the fault from sliding. This change happens at a depth of 10 kilometers or so, where the temperature reaches about 150 degrees Celsius.

Temperature also appears to control the landward, "down-dip" limit of the locked zone. At moderate temperatures the rocks show normal frictional behavior. That is, the large initial resistance to motion drops to a lesser level once the fault begins to slip. So, once sliding starts, runaway release of the stored elastic energy—an earthquake—ensues. But lower in the crust, where the temperature exceeds about 350 degrees C, the rocks lining the fault surface should behave more like a viscous fluid—faster motion meets with increasing resistance. Hence, the deeper, hotter parts of the fault are apt to creep along slowly without generating any seismic waves.

How well do these temperature limits correspond to the actual boundaries of the locked zone? My colleagues and I have tried to provide the answer by using our computer models to calculate temperatures on the subduction thrust fault. The results of that work confirmed what we had surmised. The depth on the fault where the rocks reach 350 degrees C agrees well with the down-dip limit of the locked zone as determined from the measurements of deformation.

One important question remains: Does the locked zone that we have calculated truly correspond to the source area for large subduction earthquakes? We believe it does because the vertical drop we predict for a rupture of the locked zone matches what has been observed in the buried coastal marshes. Further support comes from our efforts to apply the same techniques to other subducting margins. My colleague Kelin Wang and I, along with Makoto Yamano of the University of Tokyo, have shown that the width of the present locked zone on the Nankai margin of southwest Japan corresponds well to the rupture areas of the magnitude 8 earthquakes that struck there in the 1940s. So we can be confident that our models for Cascadia are telling us about the kind of devastating earthquakes that will eventually strike along the North American coast.

WHEN THE BIG ONE HITS

How intensely would the ground shake at the major West Coast cities during a giant subduction earthquake? The answer rests on the exact earthquake magnitude as well as on the position of the seismic source zone. The maximum magnitude that a Cascadia earthquake could achieve depends on just how far along the coast the fault releases. Simultaneous rupture of the entire stretch from British Columbia to California would be surprising because such extended breaks have been rare anywhere in the world. Yet some evidence points to a failure of this size having occurred during the Cascadia earthquake of 1700. If the total locked zone (an area of nearly 100,000 square kilometers) releases at once, a giant earthquake of magnitude 9 could result—much larger, for example, than the catastrophic San Francisco earthquake of 1906. There have been only two events of this size ever recorded: an earthquake along the coast of Chile in 1960 and the other in southern Alaska in 1964.

Seismologists can estimate the amount of ground motion that might come from such Cascadia earthquakes in two ways. One approach is to compare the situation along North America's western coast with earthquakes that have occurred elsewhere. The other method is to use rather

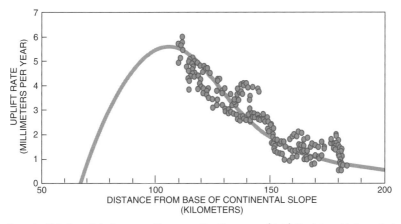

Coastal uplift in Cascadia is documented by repeated leveling surveys (*dots*). The data match theoretical predictions (*solid line*) for the bending of the North American plate and serve to locate the locked part of the subduction fault.

complicated theoretical models of the seismic rupture area and slip displacement. Either way the conclusions are similar. The next great earthquake in Cascadia will generate extremely large seismic waves lasting for as long as several minutes. After the shaking ceases, most coastal sites will be one to two meters lower and five to 10 meters seaward of where they started.

Fortunately, the locked part of the fault that would generate such earthquakes lies primarily under the continental shelf and extends little, if at all, below the coast. Hence, Vancouver, Seattle and Portland (which sit 100 to 200 kilometers inland) are subject to less severe shaking than sites near the outer western coast. Nevertheless, the seismic energy from such violent earthquakes radiates for a considerable distance, so the danger to those cities is still substantial. U.S. and Canadian residents of this Pacific coastal region who had imagined they lived on quiet ground will indeed have to learn to accept the threat of a giant earthquake upheaval happening at any moment.

FURTHER READING

SEISMIC POTENTIAL OF THE CASCADIA SUBDUCTION ZONE. Garry C. Rogers in *Nature,* Vol. 332, page 17; March 3, 1988.
CASCADIA SUBDUCTION ZONE: THE CALM BEFORE THE QUAKE? Thomas H. Heaton in *Nature,* Vol. 343, pages 511–512; February 8, 1990.

THERMAL CONSTRAINTS ON THE ZONE OF MAJOR THRUST EARTHQUAKE FAILURE: THE CASCADIA SUBDUCTION ZONE. R. B. Hyndman and K. Wang in *Journal of Geophysical Research (Solid Earth)*, Vol. 98, No. 2, pages 2039–2060; February 10, 1993.

CURRENT DEFORMATION AND THE WIDTH OF THE SEISMOGENIC ZONE OF THE NORTHERN CASCADIA SUBDUCTION THRUST. H. Dragert et al. in *Journal of Geophysical Research (Solid Earth)*, Vol. 99, No. 1, pages 653–668; January 10, 1994.

Tsunami!

FRANK I. GONZÁLEZ

ORIGINALLY PUBLISHED IN MAY 1999

The sun had set 12 minutes earlier, and twilight was waning on the northern coast of Papua New Guinea. It was July 17, 1998, and another tranquil Friday evening was drawing to a close for the men, women and children of Sissano, Arop, Warapu and other small villages on the peaceful sand spit between Sissano Lagoon and the Bismarck Sea. But deep in the earth, far beneath the wooden huts of the unsuspecting villagers, tremendous forces had strained the underlying rock for years. Now, in the space of minutes, this pent-up energy violently released as a magnitude 7.1 earthquake. At 6:49 P.M., the main shock rocked 30 kilometers (nearly 19 miles) of coastline centered on the lagoon and suddenly deformed the offshore ocean bottom. The normally flat sea surface lurched upward in response, giving birth to a fearsome tsunami.

Retired Colonel John Sanawe, who lived near the southeast end of the sandbar at Arop, survived the tsunami and later told his story to Hugh Davies of the University of Papua New Guinea. Just after the main shock struck only 20 kilometers offshore, Sanawe saw the sea rise above the horizon and then spray vertically perhaps 30 meters. Unexpected sounds— first like distant thunder, then like a nearby helicopter—gradually faded as he watched the sea slowly recede below the normal low-water mark. After four or five minutes of silence, he heard a rumble like that of a low-flying jet plane. Sanawe spotted the first tsunami wave, perhaps three or four meters high. He tried to run home, but the wave overtook him. A second, larger wave flattened the village and swept him a kilometer into a mangrove forest on the inland shore of the lagoon.

Other villagers were not so fortunate as Sanawe. Some were swept across the lagoon and impaled on the broken mangrove branches. Many more were viciously battered by debris. At least 30 survivors would lose injured limbs to gangrene. Saltwater crocodiles and wild dogs preyed on the dead before help could arrive, making it more difficult to arrive at an exact death toll. It now appears that the tsunami killed more than

2,200 villagers, including more than 230 children. Waves up to 15 meters high, which struck within 15 minutes of the main shock, had caught many coastal inhabitants unawares. Of the few villagers who knew of the tsunami hazard, those trapped on the sandbar simply had no safe place to flee.

Tsunamis such as those that pounded Papua New Guinea are the world's most powerful waves. Historical patterns of their occurrence are revealed in large databases developed by James F. Lander, Patricia A. Lockridge and their colleagues at the National Geophysical Data Center in Boulder, Colo., and Viacheslav K. Gusiakov and his associates at the Tsunami Laboratory in Novosibirsk, Russia. Most tsunamis afflict the Pacific Ocean, and 86 percent of those are the products of undersea earthquakes around the Pacific Rim, where powerful collisions of tectonic plates form highly seismic subduction zones.

Since 1990, 10 tsunamis have taken more than 4,000 lives. In all, 82 were reported worldwide—a rate much higher than the historical average of 57 a decade. The increase in tsunamis reported is due to improved global communications; the high death tolls are partly due to increases in coastal populations. My colleagues and I at the National Oceanic and Atmospheric Administration Pacific Marine Environmental Laboratory in Seattle set up an electronic-mail network as a way for researchers in distant parts of the world to help one another make faster and more accurate tsunami surveys. This Tsunami Bulletin Board, now managed by the International Tsunami Information Center, has facilitated communication among tsunami scientists since shortly after the 1992 Nicaragua tsunami.

Disasters similar to those in Nicaragua and Papua New Guinea have wreaked havoc in Hawaii and Alaska in the past, but most tsunami researchers had long believed that the U.S. West Coast was relatively safe from the most devastating events. New evidence now suggests that earthquakes may give birth to large tsunamis every 300 to 700 years along the Cascadia subduction zone, an area off the Pacific Northwest coast where a crustal plate carrying part of the Pacific Ocean is diving under North America. A clear reminder of this particular threat occurred in April 1992, when a magnitude 7.1 earthquake at the southern end of the subduction zone generated a small tsunami near Cape Mendocino, Calif. This event served as the wake-up call that has driven the development of the first systematic national effort to prepare for dangerous tsunamis before they strike. The Pacific Marine Environmental Laboratory is playing a key research and management role in this endeavor.

THE PHYSICS OF TSUNAMIS

To understand tsunamis, it is first helpful to distinguish them from wind-generated waves or tides. Breezes blowing across the ocean crinkle the surface into relatively short waves that create currents restricted to a shallow layer; a scuba diver, for example, might easily swim deep enough to find calm water. Strong gales are able to whip up waves 30 meters or higher in the open ocean, but even these do not move deep water.

.Tides, which sweep around the globe twice a day, do produce currents that reach the ocean bottom—just as tsunamis do. Unlike true tidal waves, however, tsunamis are not generated by the gravitational pull of the moon or sun. A tsunami is produced impulsively by an undersea earthquake or, much less frequently, by volcanic eruptions, meteorite impacts or underwater landslides. With speeds that can exceed 700 kilometers per hour in the deep ocean, a tsunami wave could easily keep pace with a Boeing 747. Despite its high speed, a tsunami is not dangerous in deep water. A single wave is less than a few meters high, and its length can extend more than 750 kilometers in the open ocean. This creates a sea-surface slope so gentle that the wave usually passes unnoticed in deep water. In fact, the Japanese word *tsu-nami* translates literally as "harbor wave," perhaps because a tsunami can speed silently and undetected across the ocean, then unexpectedly arise as destructively high waves in shallow coastal waters.

A powerful tsunami also has a very long reach: it can transport destructive energy from its source to coastlines thousands of kilometers away. Hawaii, because of its midocean location, is especially vulnerable to such Pacific-wide tsunamis. Twelve damaging tsunamis have struck Hawaii since 1895. In the most destructive, 159 people died there in 1946 from killer waves generated almost 3,700 kilometers away in Alaska's Aleutian Islands. Such remote-source tsunamis can strike unexpectedly, but local-source tsunamis—as in the case of last year's Papua New Guinea disaster—can be especially devastating. Lander has estimated that more than 90 percent of all fatalities occur within about 200 kilometers of the source. As an extreme example, it is believed that a tsunami killed more than 30,000 people within 120 kilometers of the catastrophic eruption of Krakatoa volcano in the Sunda Straits of Indonesia in 1883. That explosion generated waves as high as a 12-story building.

Regardless of their origin, tsunamis evolve through three overlapping but quite distinct physical processes: generation by any force that disturbs the water column, propagation from deeper water near the source to shallow coastal areas and, finally, inundation of dry land. Of these, the

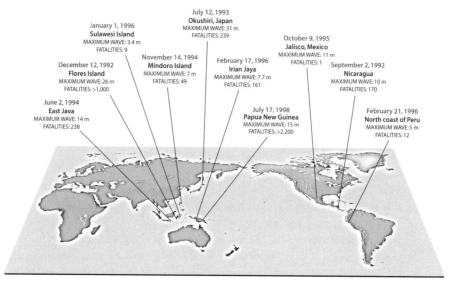

July 12, 1993
Okushiri, Japan
MAXIMUM WAVE: 31 m
FATALITIES: 239

January 1, 1996
Sulawesi Island
MAXIMUM WAVE: 3.4 m
FATALITIES: 9

October 9, 1995
Jalisco, Mexico
MAXIMUM WAVE: 11 m
FATALITIES: 1

November 14, 1994
Mindoro Island
MAXIMUM WAVE: 7 m
FATALITIES: 49

December 12, 1992
Flores Island
MAXIMUM WAVE: 26 m
FATALITIES: >1,000

February 17, 1996
Irian Jaya
MAXIMUM WAVE: 7.7 m
FATALITIES: 161

September 2, 1992
Nicaragua
MAXIMUM WAVE: 10 m
FATALITIES: 170

June 2, 1994
East Java
MAXIMUM WAVE: 14 m
FATALITIES: 238

July 17, 1998
Papua New Guinea
MAXIMUM WAVE: 15 m
FATALITIES: >2,200

February 21, 1996
North coast of Peru
MAXIMUM WAVE: 5 m
FATALITIES: 12

Ten destructive tsunamis have claimed more than 4,000 lives since 1990. Last year's Papua New Guinea disaster is the most recent in this string of killer waves generated by earthquakes along colliding tectonic plates of the Pacific Rim.

propagation phase is best understood, whereas generation and inundation are more difficult to model with computer simulations. Accurate simulations are important in predicting where future remote-source tsunamis will strike and in guiding disaster surveys and rescue efforts, which must concentrate their resources on regions believed to be hardest hit.

Generation is the process by which a seafloor disturbance, such as movement along a fault, reshapes the sea surface into a tsunami. Modelers assume that this sea-surface displacement is identical to that of the ocean bottom, but direct measurements of seafloor motion have never been available (and may never be). Instead researchers use an idealized model of the quake: they assume that the crustal plates slip past one another along a simple, rectangular plane inside the earth. Even then, predicting the tsunami's initial height requires at least 10 descriptive parameters, including the amount of slip on each side of the imaginary plane and its length and width. As modelers scramble to guide tsunami survey teams immediately after an earthquake, only the orientation of the assumed fault plane and the quake's location, magnitude and depth can be interpreted from the seismic data alone. All other parameters must be estimated. As a consequence, this first simulation frequently underestimates inundation, sometimes by factors of 5 or 10.

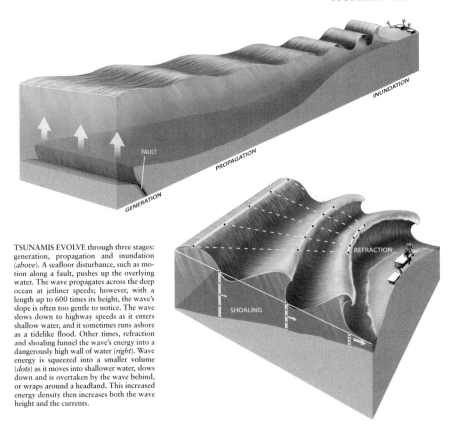

TSUNAMIS EVOLVE through three stages: generation, propagation and inundation (*above*). A seafloor disturbance, such as motion along a fault, pushes up the overlying water. The wave propagates across the deep ocean at jetliner speeds; however, with a length up to 600 times its height, the wave's slope is often too gentle to notice. The wave slows down to highway speeds as it enters shallow water, and it sometimes runs ashore as a tidelike flood. Other times, refraction and shoaling funnel the wave's energy into a dangerously high wall of water (*right*). Wave energy is squeezed into a smaller volume (*dots*) as it moves into shallower water, slows down and is overtaken by the wave behind, or wraps around a headland. This increased energy density then increases both the wave height and the currents.

Low inundation estimates can signify that the initial tsunami height was also understated because the single-plane fault model distributes seismic energy over too large an area. Analyses of seismic data cannot resolve energy distribution patterns any shorter than the seismic waves themselves, which extend for several hundred kilometers. But long after the tsunami strikes land, modelers can work backward from records of run-up and additional earthquake data to refine the tsunami's initial height. For example, months of aftershocks eventually reveal patterns of seismic energy that are concentrated in regions much smaller than the original, single-plane fault model assumed. When seismic energy is focused in a smaller area, the vertical motion of the seafloor—and therefore the initial tsunami height—is greater. Satisfactory simulations are achieved only after months of labor-intensive work, but every simulation that matches the real disaster improves scientists' ability to make better predictions.

Propagation of the tsunami transports seismic energy away from the earthquake site through undulations of the water, just as shaking moves the energy through the earth. At this point, the wave height is so small compared with both the wavelength and the water depth that researchers apply linear wave theory, which assumes that the height itself does not affect the wave's behavior. The theory predicts that the deeper the water and the longer the wave, the faster the tsunami. This dependence of wave speed on water depth means that refraction by bumps and grooves on the seafloor can shift the wave's direction, especially as it travels into shallow water. In particular, wave fronts tend to align parallel to the shoreline so that they wrap around a protruding headland before smashing into it with greatly focused incident energy. At the same time, each individual wave must also slow down because of the decreasing water depth, so they begin to overtake one another, decreasing the distance between them in a process called shoaling. Refraction and shoaling squeeze the same amount of energy into a smaller volume of water, creating higher waves and faster currents.

The last stage of evolution, inundation and run-up, in which a tsunami may run ashore as a breaking wave, a wall of water or a tidelike flood, is perhaps the most difficult to model. The wave height is now so large that linear theory fails to describe the complicated interaction between the water and the shoreline. Vertical run-up can reach tens of meters, but it typically takes only two to three meters to cause damage. Horizontal inundation, if unimpeded by coastal cliffs or other steep topography, can penetrate hundreds of meters inland. Both kinds of flooding are aided and abetted by the typical crustal displacement of a subduction zone earthquake, which lifts the offshore ocean bottom and lowers the land along the coast. This type of displacement propagates waves seaward with a leading crest and landward with a leading trough (the reason a receding sea sometimes precedes a tsunami). Not only does the near-shore subsidence facilitate tsunami penetration inland but, according to recent studies by Raissa Mazova of the Nizhny Novgorod State Technical University in Russia and by Costas Synolakis of the University of Southern California, both theoretical predictions and field surveys indicate that coastal run-up and inundation will be greater if the trough of the leading wave precedes the crest.

TSUNAMI THREATS

Predicting where a tsunami may strike helps to save lives and property only if coastal inhabitants recognize the threat and respond appropriately.

More than a quarter of all reliably reported Pacific tsunamis since 1895 originated near Japan. This is not surprising, because Japan is precariously situated near the colliding margins of four tectonic plates. Recognizing the recurring threat, the Japanese have invested heavily over the years in tsunami hazard mitigation, including comprehensive educational and public outreach programs, an effective warning system, shoreline barrier forests, seawalls and other coastal fortifications.

On the night of July 12, 1993, their preparations faced a brutal test. A magnitude 7.8 earthquake in the Sea of Japan generated a tsunami that struck various parts of the small island of Okushiri. Five minutes after the main shock the Japan Meteorological Agency issued a warning over television and radio that a major tsunami was on its way. By then, 10- to 20-meter waves had struck the coastline nearest the source, claiming a number of victims before they could flee. In Aonae, a small fishing village on the island's southern peninsula, many of the 1,600 townspeople fled to high ground as soon as they felt the main shock. A few minutes later tsunami waves five to 10 meters high ravaged hundreds of their homes and businesses and swept them out to sea. More than 200 lives were lost in this disaster, but quick response saved many more.

Over the past century in Japan, approximately 15 percent of 150 tsunamis were damaging or fatal. That track record is much better than the tally in countries with few or no community education programs in place. For example, more than half of the 34 tsunamis that struck Indonesia in the past 100 years were damaging or fatal. Interviews conducted after the 1992 Flores Island tsunami that killed more than 1,000 people indicated that most coastal residents did not recognize the earthquake as the natural warning of a possible tsunami and did not flee inland. Similarly, Papua New Guinea residents were tragically uninformed, sending the number of casualties from last year's disaster higher than expected for a tsunami of that size. A large quake in 1907 evidently lowered the area that is now Sissano Lagoon, but any resulting tsunami was too small and too long ago to imprint a community memory. When the earthquake struck last year, some people actually walked to the coast to investigate the disturbance, thus sealing their fate.

Scientists have learned a great deal from recent tsunamis, but centuries-old waves continue to yield valuable insights. Lander and his colleagues have described more than 200 tsunamis known to have affected the U.S. since the time of the first written records in Alaska and the Caribbean during the early 1700s and in Hawaii and along the West Coast later that century. Total damage is estimated at half a billion dollars and 470 casualties, primarily in Alaska and Hawaii. An immediate threat to those states and the West Coast is the Alaska-Aleutian subduction zone. Included in this region's history of large, tsunami-generating earthquakes are two disasters that drove the establishment of the country's only two tsunami warning centers. The probability of a magnitude 7.4 or greater earthquake occurring somewhere in this zone before 2008 is estimated to be 84 percent.

Another major threat, unrevealed by the written records, lurks off the coasts of Washington State, Oregon and northern California—the Cascadia subduction zone. Brian F. Atwater of the U.S. Geological Survey has identified sand and gravel deposits that he hypothesized were carried inland from the Washington coast by tsunamis born of Cascadia quakes. Recent events support this theory. The Nicaragua tsunami was notable for the amount of sand it transported inland, and researchers have documented similar deposits at inundation sites in Flores, Okushiri, Papua New Guinea and elsewhere.

At least one segment of the Cascadia subduction zone may be approaching the end of a seismic cycle that culminates in an earthquake and destructive tsunami [see "Giant Earthquakes of the Pacific Northwest," by Roy D. Hyndman; *Scientific American,* December 1995]. The earthquake danger is believed to be comparable to that in southern California—about a 35 percent probability of occurrence before 2045. Finally, the 1992 Cape Mendocino earthquake and tsunami was a clear reminder that the Cascadia subduction zone can unleash local tsunamis that strike the coast within minutes.

GETTING READY IN THE U.S.

Hard on the heels of the surprising Cape Mendocino tsunami, the Federal Emergency Management Agency (FEMA) and NOAA funded an earthquake scenario study of northern California and the production of tsunami inundation maps for Eureka and Crescent City in that state. The resulting "all hazards" map was the first of its kind for the U.S. It delineates areas susceptible to tsunami flooding, earthquake-shaking intensity, liquefaction and landslides. Researchers then tackled the possible effects of a great Cascadia subduction zone earthquake and tsunami. About 300,000 people live or work in nearby coastal regions, and at least as many tourists travel through these areas every year. Local tsunami waves could strike communities within minutes of a big quake, leaving little or no time to issue formal warnings. What is more, a Cascadia-born tsunami disaster could cost the region between \$1.25 billion and \$6.25 billion, a conservative estimate considering the 1993 Okushiri disaster.

Clarification of the threat from the Cascadia subduction zone and the many well-reported tsunami disasters of this decade have stimulated a systematic effort to examine the tsunami hazard in the U.S. In 1997 Congress provided \$2.3 million to establish the National Tsunami Hazard Mitigation Program. Alaska, California, Hawaii, Oregon and Washington formed a partnership with NOAA, FEMA and the USGS to tackle the threat of both local- and remote-source tsunamis. The partnership focuses on three interlocking activities: assessing the threat to specific coastal areas; improving early detection of tsunamis and their potential danger; and educating communities to ensure an appropriate response when a tsunami strikes.

The threat to specific coastal areas can be assessed by means of tsunami inundation maps such as those designed for Eureka and Crescent City using state-of-the-art computer modeling. These maps provide critical guidance to local emergency planners charged with identifying evacuation routes. Only Hawaii has systematically developed such maps over the years. To date, three Oregon communities have received maps, six additional maps are in progress in Oregon, Washington and California, and three maps are planned for Alaska.

Rapid, reliable confirmation of the existence of a potentially dangerous tsunami is essential to officials responsible for sounding alarms. Coastal tide gauges have been specially modified to measure tsunamis, and a major upgrade of the seismic network will soon provide more rapid and more complete reports on the nature of the earthquake. These instruments are essential to the warning system, but seismometers measure earthquakes, not tsunamis. And although tide gauges spot tsunamis close to shore, they cannot measure tsunami energy propagating toward a distant coastline. As a consequence, an unacceptable 75 percent false-alarm rate has prevailed since the 1950s. These incidents are expensive, undermine the credibility of the warning system, and place citizens at risk during the evacuation. A false alarm that triggered the evacuation of Honolulu on May 7, 1986, cost Hawaii more than \$30 million in lost salaries and business revenues.

NOAA is therefore developing a network of six deep-ocean reporting stations that can track tsunamis and report them in real time, a project known as Deep-Ocean Assessment and Reporting of Tsunamis (DART). Scientists have completed testing of prototype systems and expect the network to be operating reliably in two years. The rationale for this type of warning system is simple: if an earthquake strikes off the coast of Alaska while you're lying on a Hawaiian beach, what you really want to have between you and the quake's epicenter is a DART system. Here's why:

Seismometers staked out around the Pacific Rim can almost instantly pinpoint a big Alaskan quake's location. In the next moment, complex computer programs can predict how long a triggered tsunami would take to reach Hawaii, even though there is not yet evidence a wave exists. After some minutes, tide gauges scattered along the coastlines may detect a tsunami. But the only way to be sure whether a dangerous wave is headed toward a distant coast is to place tsunami detectors in its path and track it across the open ocean.

Conceptually, the idea of such a real-time reporting network is straightforward; however, formidable technological and logistical challenges have held up implementation until now. The DART systems depend on bottom pressure recorders that Hugh B. Milburn, Alex Nakamura, Eddie N. Bernard and I have been perfecting over the past decade at the Pacific Marine Environmental Laboratory. As the crest of a tsunami wave passes by, the bottom recorder detects the increased pressure from the additional volume of overlying water. Even 6,000 meters deep, the sensitive instrument can detect a tsunami no higher than a single centimeter. Ship and storm waves are not detected, because their length is short and, as with currents, changes in pressure are not transmitted all the way to the ocean bottom. We placed the first recorders on the north Pacific seafloor in 1986 and have been using them to record tsunamis ever since. The records cannot be accessed, however, until the instruments are retrieved.

Ideally, when the bottom recorders detect a tsunami, acoustic chirps will transmit the measurements to a car-size buoy at the ocean surface, which will then relay the information to a ground station via satellite. The surface buoy systems, the satellite relay technology and the bottom recorders have proved themselves at numerous deep-ocean stations, including an array of 70 weather buoys set up along the equator to track El Niño, the oceanographic phenomenon so infamous for its effect on world climate. The biggest challenge has been developing a reliable acoustic transmission system. Over the past three years, four prototype DART systems have been deployed, worked for a time, then failed. Design improvements to a second-generation system have refined communication between the bottom recorders and the buoys.

In the next two years, our laboratory plans to establish five stations spread across the north Pacific from the west Aleutians to Oregon and a sixth sited on the equator to intercept tsunamis generated off South America. More buoys would reduce the possibility that tsunami waves might

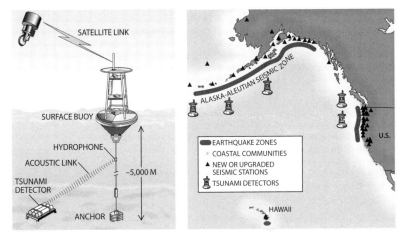

Deep-ocean tsunami detectors (*left*) and a major upgrade of existing earthquake monitoring networks (*triangles on map*)—both scheduled for installation within two years—lead the U.S. effort to take the surprise out of tsunami attacks. The deep-ocean detectors depend on high-tech sensors stationed on the seafloor. When one of these instruments senses a tsunami wave overhead, it will send acoustic signals to a buoy at the surface, which will then relay the warning via satellite to the officials who are responsible for sounding an alarm.

sneak between them, but the current budget limits the number that NOAA can afford. This is where detailed computer simulations become invaluable. Combined with the buoy measurements, the simulations will provide more accurate predictions to guide officials who may have only a few minutes to decide whether to sound an alarm.

Even the most reliable warning is ineffective if people do not respond appropriately. Community education is thus perhaps the most important aspect of the national mitigation program's threefold mission. Each state is identifying coordinators who will provide information and guidance to community emergency managers during tsunami disasters. Interstate co-ordination is also crucial to public safety because U.S. citizens are highly mobile, and procedures must be compatible from state to state. Standard tsunami signage has already been put in place along many coastlines.

Tsunami researchers and emergency response officials agree that future destructive tsunamis are inevitable and technology alone cannot save lives. Coastal inhabitants must be able to recognize the signs of a possible tsunami—such as strong, prolonged ground shaking—and know that they should seek higher ground immediately. Coastal communities need inundation maps that identify far in advance what areas are likely to be flooded so that they can lay out evacuation routes. The proactive enterprise now under way in the U.S. will surely upgrade tsunami prediction for a much larger region of the Pacific. All of these efforts are essential to the overriding goal of avoiding tragedies such as those in Papua New Guinea, Nicaragua and elsewhere.

FURTHER READING

UNITED STATES TSUNAMIS (INCLUDING UNITED STATES POSSESSIONS): 1690–1988. James F. Lander and Patricia A. Lockridge, NOAA/National Geophysical Data Center, Publication 41–42, 1989.

THE CAPE MENDOCINO TSUNAMI. F. I. González and E. N. Bernard in *Earthquakes and Volcanoes,* Vol. 23, No. 3, pages 135–138; 1992.

TSUNAMI! Walter C. Dudley and Min Lee. University of Hawaii Press, 1998.

Additional information on tsunamis can be found at http://www.pmel.noaa/tsunami/ on the World Wide Web.

The Threat of Silent Earthquakes

PETER CERVELLI

ORIGINALLY PUBLISHED IN MARCH 2004

In early November 2000 the Big Island of Hawaii experienced its largest earthquake in more than a decade. Some 2,000 cubic kilometers of the southern slope of Kilauea volcano lurched toward the ocean, releasing the energy of a magnitude 5.7 shock. Part of that motion took place under an area where thousands of people stop every day to catch a glimpse of one of the island's most spectacular lava flows. Yet when the earthquake struck, no one noticed—not even seismologists.

How could such a notable event be overlooked? As it turns out, quaking is not an intrinsic part of all earthquakes. The event on Kilauea was one of the first unambiguous records of a so-called silent earthquake, a type of massive earth movement unknown to science until just a few years ago. Indeed, I would never have discovered this quake if my colleagues at the U.S. Geological Survey's Hawaiian Volcano Observatory had not already been using a network of sensitive instruments to monitor the volcano's activity. When I finally noticed that Kilauea's south flank had shifted 10 centimeters along an underground fault, I also saw that this movement had taken nearly 36 hours—a turtle's pace for an earthquake. In a typical tremor, opposite sides of the fault rocket past each other in a matter of seconds—quickly enough to create the seismic waves that cause the ground to rumble and shake.

But just because an earthquake happens slowly and quietly does not make it insignificant. My co-investigators and I realized immediately that Kilauea's silent earthquake could be a harbinger of disaster. If that same large body of rock and debris were to gain momentum and take the form of a gigantic landslide—separating itself from the rest of the volcano and sliding rapidly into the sea—the consequences would be devastating. The collapsing material would push seawater into towering tsunami waves that could threaten coastal cities along the entire Pacific Rim. Such catastrophic flank failure, as geologists call it, is a potential threat around many island volcanoes worldwide.

UNEXPECTED STIR

Fortunately, the discovery of silent earthquakes is revealing more good news than bad. The chances of catastrophic flank failure are slim, and the instruments that record silent earthquakes might make early warnings possible. New evidence for conditions that might trigger silent slip suggests bold strategies for preventing flank collapse. Occurrences of silent earthquakes are also being reported in areas where flank failure is not an issue. There silent earthquakes are inspiring ways to improve forecasts of their ground-shaking counterparts.

The discovery of silent earthquakes and their link to catastrophic flank collapse was a by-product of efforts to study other potential natural hazards. Destructive earthquakes and volcanoes are a concern in Japan and the U.S. Pacific Northwest, where tectonic plates constantly plunge deep into the earth along what are called subduction zones. Beginning in the early 1990s, geologists began deploying large networks of continuously recording Global Positioning System (GPS) receivers in these regions and along the slopes of active volcanoes, such as Kilauea. By receiving signals from a constellation of more than 30 navigational satellites, these instruments can measure their own positions on the planet's surface at any given time to within a few millimeters.

The scientists who deployed these GPS receivers expected to see both the slow, relentless motion of the planet's shell of tectonic plates and the relatively quick movements that earthquakes and volcanoes trigger. It came as some surprise when these instruments detected small ground movements that were not associated with any known earthquake or eruption. When researchers plotted the ground movements on a map, the pattern that resulted very much resembled one characteristic of fault movement. In other words, all the GPS stations on one side of a given fault moved several centimeters in the same general direction. This pattern would have been no surprise if it had taken a year or longer to form. In that case, scientists would have known that a slow and steady process called fault creep was responsible. But at rates of up to centimeters a day, the mystery events were hundreds of times as fast as that. Beyond their relative speediness, these silent earthquakes shared another attribute with their noisy counterparts that distinguished them from fault creep: they are not steady processes but instead are discrete events that begin and end suddenly.

That sudden beginning, when it takes place on the slopes of a volcanic island, creates concern about a possible catastrophic flank event. Most typical earthquakes happen along faults that have built-in brakes: motion stops once the stress is relieved between the two chunks of earth that are trying to move past each other. But activity may not stop if gravity becomes the primary driver. In the worst-case scenario, the section of the volcano lying above the fault becomes so unstable that once slip starts, gravity pulls the entire mountainside downhill until it disintegrates into a pile of debris on the ocean floor.

The slopes of volcanoes such as Kilauea become steep and vulnerable to this kind of collapse when the lava from repeated eruptions builds them up more rapidly than they can erode away. Discovering the silent earthquake on Kilauea suggests that the volcano's south flank is on the move—perhaps on its way to eventual obliteration.

For now, friction along the fault is acting like an emergency brake. But gravity has won out in many other instances in the past. Scientists have long seen evidence of ancient collapses in sonar images of giant debris fields in the shallow waters surrounding volcanic islands around the world, including Majorca in the Mediterranean Sea and the Canary Islands in the Atlantic Ocean. In the Hawaiian Islands, geologists have found more than 25 individual collapses that have occurred over the past five million years—the blink of an eye in geologic time.

In a typical slide, the volume of material that enters the ocean is hundreds of times as great as the section of Mount St. Helens that blew apart during the 1980 eruption—more than enough to have triggered immense tsunamis. On the Hawaiian island of Lanai, for instance, geologists discovered evidence of wave action, including abundant marine shell fragments, at elevations of 325 meters. Gary M. McMurtry of the University of Hawaii at Manoa and his colleagues conclude that the most likely way the shells could have reached such a lofty location was within the waves of a tsunami that attained the astonishing height of 300 meters along some Hawaiian coastlines. Most of the tallest waves recorded in modern times were no more than one tenth that size.

PREPARING FOR THE WORST

As frightening as such an event may sound, this hazard must be understood in the proper context. Catastrophic failure of volcanic slopes is very rare on a human timescale—though far more common than the potential for a large asteroid or comet to have a damaging collision with the earth. Collapses large enough to generate a tsunami occur somewhere in the Hawaiian Islands only about once every 100,000 years. Some scientists

THE MECHANICS OF SILENT EARTHQUAKES

PERCOLATING WATER may trigger silent earthquakes if it finds a way into a vulnerable fault. Highly pressurized by the burden of overlying rock, water can push apart the two sides of the fault (*inset*), making it easier for them to slip past each other (*red arrows*). This kind of silent slip can occur within subduction zones and volcanic islands.

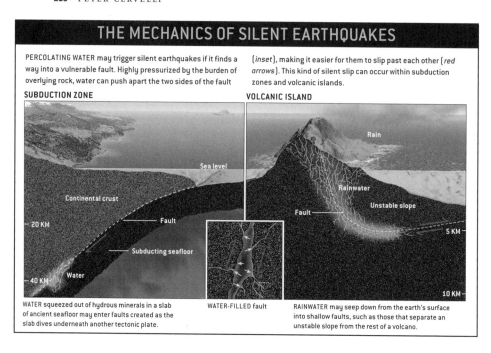

WATER squeezed out of hydrous minerals in a slab of ancient seafloor may enter faults created as the slab dives underneath another tectonic plate.

WATER-FILLED fault

RAINWATER may seep down from the earth's surface into shallow faults, such as those that separate an unstable slope from the rest of a volcano.

estimate that such events occur worldwide once every 10,000 years. Because the hazard is extremely destructive when it does happen, many scientists agree that it is worth preparing for.

To detect deformation within unstable volcanic islands, networks of continuous GPS receivers are beginning to be deployed on Réunion Island in the Indian Ocean, on Fogo in the Cape Verde Islands, and throughout the Galápagos archipelago, among others. Kilauea's network of more than 20 GPS stations, for example, has already revealed that the volcano experiences creep, silent earthquakes as well as large, destructive typical earthquakes. Some scientists propose, however, that Kilauea may currently be protected from catastrophic collapse by several underwater piles of mud and rock—probably debris from old flank collapses—that are buttressing its south flank. New discoveries about the way Kilauea is slipping can be easily generalized to other island volcanoes that may not have similar buttressing structures.

Whatever the specific circumstances for an island, the transition from silent slip to abrupt collapse would involve a sudden acceleration of the mobile slope. In the worst case, this acceleration would proceed immediately to breakneck velocities, leaving no chance for early detection and warning. In the best case, the acceleration would occur in fits and starts, in a cascade of silent earthquakes slowly escalating into regular earthquakes, and then on to catastrophe. A continuous GPS network could easily detect this fitful acceleration, well before ground-shaking earthquakes began to occur and, with luck, in plenty of time for a useful tsunami warning.

If the collapse were big enough, however, a few hours' or even days' warning might come as little comfort because it would be so difficult at that point to evacuate everyone. This problem raises the question of whether authorities might ever implement preventive measures. The problem of stabilizing the teetering flanks of oceanic volcanoes is solvable—in principle. In practice, however, the effort required would be immense. Consider simple brute force. If enough rock were removed from the upper reaches of an unstable volcanic flank, then the gravitational potential energy that is driving the system toward collapse would disappear for at least several hundred thousand years. A second possible method—lowering an unstable flank slowly through a series of small earthquakes—would be much cheaper but fraught with geologic unknowns and potential dangers. To do so, scientists could conceivably harness as a tool to prevent collapse the very thing that may be currently driving silent earthquakes on Kilauea.

Nine days before the most recent silent earthquake on Kilauea, a torrential rainstorm dropped nearly a meter of water on the volcano in less than 36 hours. Geologists have long known that water leaking into faults can trigger earthquakes, and nine days is about the same amount of time that they estimate it takes water to work its way down through cracks and pores in Kilauea's fractured basaltic rock to a depth of five kilometers—where the silent earthquake occurred. My colleagues and I suspect that the burden of the overlying rock pressurized the rainwater, forcing the sides of the fault apart and making it much easier for them to slip past each other.

This discovery lends credence to the controversial idea of forcefully injecting water or steam into faults at the base of an unstable flank to trigger the stress-relieving earthquakes needed to let it down slowly. This kind of human-induced slip happens at very small scales all the time at geothermal plants and other locations where water is pumped into the earth. But when it comes to volcanoes, the extreme difficulty lies in putting the right amount of fluid in the right place so as not to inadvertently generate the very collapse that is meant to be avoided. Some geophysicists considered this strategy as a way to relieve stress along California's infamous San Andreas fault, but they ultimately abandoned the idea for fear that it would create more problems than it would solve.

WEDGES OF WATER

Apart from calling attention to the phenomenon of catastrophic collapse of the flank of a volcano, the discovery of silent earthquakes is forcing scientists to reconsider various aspects of fault motion—including seismic hazard assessments. In the U.S. Pacific Northwest, investigators have observed many silent earthquakes along the enormous Cascadia fault zone between the North American plate and the subducting Juan de Fuca plate. One curious feature of these silent earthquakes is that they happen at regular intervals—so regular, in fact, that scientists are now predicting their occurrence successfully.

This predictability most likely stems from the fact that water flowing from below subduction zones may exert significant control over when and where these faults slip silently. As the subducting plate sinks deeper into the earth, it encounters higher and higher temperatures and pressures, which release the significant amount of water trapped in water-rich minerals that exist within the slab. The silent earthquakes may then take place when a batch of fluid from the slab is working its way up—as the fluid passes, it will unclamp the fault zone a little bit, perhaps allowing some slow slip.

What is more, Garry Rogers and Herb Dragert of the Geological Survey of Canada reported last June that these silent tremors might even serve as precursors to some of the region's large, ground-shaking shocks. Because the slow slips occur deep and at discrete intervals, they regulate the rate at which stress accumulates on the shallower part of the fault zone, which moves in fits and starts. In this shallow, locked segment of the fault, it usually takes years or even centuries to amass the stress required to set off a major shock. Rogers and Dragert suggest, however, that silent slip may dramatically hasten this stress buildup, thereby increasing the risk of a regular earthquake in the weeks and months after a silent one.

Silent earthquakes are forcing scientists to rethink seismic forecasts in other parts of the world as well. Regions of Japan near several so-called seismic gaps—areas where fewer than expected regular earthquakes occur in an otherwise seismically active region—are thought to be overdue for a destructive shock. But if silent slip has been relieving stress along these faults without scientists realizing it, then the degree of danger may actually be less than they think. Likewise, if silent slip is discovered along faults that were considered inactive up to now, these structures will need careful evaluation to determine whether they are also capable of destructive earthquakes.

If future study reveals silent earthquakes to be a common feature of most large faults, then scientists will be forced to revisit long-held doctrines about all earthquakes. The observation of many different speeds of fault slip poses a real challenge to theorists trying to explain the faulting process with fundamental physical laws, for example. It is now believed that the number and sizes of observed earthquakes can be explained with a fairly simple friction law. But can this law also account for silent earthquakes? So far no definitive answer has been found, but research continues.

Silent earthquakes are only just beginning to enter the public lexicon. These subtle events portend an exponential increase in our understanding of the how and why of fault slip. The importance of deciphering fault slip is difficult to overstate because when faults slip quickly, they can cause immense damage, sometimes at a great distance from the source. The existence of silent earthquakes gives scientists a completely new angle on the slip process by permitting the detailed study of fault zones through every stage of their movement.

FURTHER READING

Sudden Aseismic Fault Slip on the South Flank of Kilauea Volcano, Hawaii. Peter Cervelli, Paul Segall, Kaj Johnson, Michael Lisowski and Asta Miklius in *Nature,* Vol. 415, pages 1014–1017; February 28, 2002.

Episodic Tremor and Slip on the Cascadia Subduction Zone: The Chatter of Silent Slip. Garry Rogers and Herb Dragert in *Science,* Vol. 300, pages 1942–1943; June 20, 2003.

Giant Landslides, Mega-Tsunamis, and Paleo-Sea Level in the Hawaiian Islands. G. M. McMurtry, P. Watts, G. J. Fryer, J. R. Smith and F. Imamura in *Marine Geology.* Available online at www.sciencedirect.com/science/journal/00253227

Visit the U.S. Geological Survey Hawaiian Volcano Observatory at http://hvo.wr.usgs.gov

The Coming Climate

THOMAS R. KARL, NEVILLE NICHOLLS AND JONATHAN GREGORY

ORIGINALLY PUBLISHED IN MAY 1997

Human beings have in recent years discovered that they may have succeeded in achieving a momentous but rather unwanted accomplishment. Because of our numbers and our technology, it now seems likely that we have begun altering the climate of our planet.

Climatologists are confident that over the past century, the global average temperature has increased by about half a degree Celsius. This warming is thought to be at least partly the result of human activity, such as the burning of fossil fuels in electric power plants and automobiles. Moreover, because populations, national economies and the use of technology are all growing, the global average temperature is expected to continue increasing, by an additional 1.0 to 3.5 degrees C by the year 2100.

Such warming is just one of many consequences that climate change can have. Nevertheless, the ways that warming might affect the planet's environment—and, therefore, its life—are among the most compelling issues in earth science. Unfortunately, they are also among the most difficult to predict. The effects will be complex and vary considerably from place to place. Of particular interest are the changes in regional climate and local weather and especially extreme events—record temperatures, heat waves, very heavy rainfall, or drought, for example—which could very well have staggering effects on societies, agriculture and ecosystems.

Based on studies of how the earth's weather has changed over the past century as global temperatures edged upward, as well as on sophisticated computer models of climate, it now seems probable that warming will accompany changes in regional weather. For example, longer and more intense heat waves—a likely consequence of an increase in either the mean temperature or in the variability of daily temperatures—would result in public health threats and even unprecedented levels of mortality, as well as in such costly inconveniences as road buckling and high cooling loads, the latter possibly leading to electrical brownouts or blackouts.

Climate change would also affect the patterns of rainfall and other precipitation, with some areas getting more and others less, changing global patterns and occurrences of droughts and floods. Similarly, increased variability and extremes in precipitation can exacerbate existing problems in water quality and sewage treatment and in erosion and urban stormwater routing, among others. Such possibilities underscore the need to understand the consequences of humankind's effect on global climate.

TWO PRONGS

Researchers have two main—and complementary—methods of investigating these climate changes. Detailed meteorological records go back about a century, which coincides with the period during which the global average temperature increased by half a degree. By examining these measurements and records, climatologists are beginning to get a picture of how and where extremes of weather and climate have occurred.

It is the relation between these extremes and the overall temperature increase that really interests scientists. This is where another critical research tool—global ocean-atmosphere climate models—comes in. These high-performance computer programs simulate the important processes of the atmosphere and oceans, giving researchers insights into the links between human activities and major weather and climate events.

The combustion of fossil fuels, for example, increases the concentration in the atmosphere of certain greenhouse gases, the fundamental agents of the global warming that may be attributable to humans. These gases, which include carbon dioxide, methane, ozone, halocarbons and nitrous oxide, let in sunlight but tend to insulate the planet against the loss of heat, not unlike the glass of a greenhouse. Thus, a higher concentration means a warmer climate.

Of all the human-caused (anthropogenic) greenhouse gases, carbon dioxide has by far the greatest impact on the global heat budget (calculated as the amount of heat absorbed by the planet less the amount radiated back into space). Contributing to carbon dioxide's greenhouse potency is its persistence: as much as 40 percent of it tends to remain in the atmosphere for centuries. Accumulation of atmospheric carbon dioxide is promoted not only by combustion but also by tropical deforestation.

The second most influential human-caused effect on the earth's radiation budget is probably that of aerosols, which are minute solid particles, sometimes covered by a liquid film, finely dispersed in the atmosphere. They, too, are produced by combustion, but they also come from natural sources, primarily volcanoes. By blocking or reflecting light, aerosols tend to mitigate global warming on regional and global scales. In contrast to carbon dioxide, aerosols have short atmospheric residence times (less than a week) and consequently are concentrated near their sources. At present, scientists are less certain about the radiative effects of aerosols than those of greenhouse gases.

By taking increases in greenhouse gases into account, global ocean-atmosphere climate models can provide some general indications of what we might anticipate regarding changes in weather events and extremes. Unfortunately, however, the capabilities of even the fastest computers and our limited understanding of the linkages among various atmospheric, climatic, terrestrial and oceanic phenomena limit our ability to model important details on the scales at which they occur. For example, clouds are of great significance in the atmospheric heat budget, but the physical processes that form clouds and determine their characteristics operate on scales too small to be accounted for directly in global-scale simulations.

HOW HOT, AND HOW OFTEN?

The deficiencies in computer models become rather apparent in efforts to reproduce or predict the frequency of climate and weather extremes of all kinds. Of these extremes, temperature is one of the most closely studied, because of its effect on humanity, through health and mortality, as well as cooling loads and other factors. Fortunately, researchers have been able to garner some insights about these extremes by analyzing decades of weather data. For statistical reasons, even slight increases in the average temperature can result in big jumps in the number of very warm days [*see top illustration on next page*].

One of the reasons temperature extremes are so difficult to model is that they are particularly sensitive to unusual circulation patterns and air masses, which can occasionally cause them to follow a trend in the direction opposite that of the mean temperature. For example, in the former Soviet Union, the annual extreme minimum temperature has increased by a degree and a half, whereas the annual extreme maximum showed no change.

The National Climatic Data Center, which is part of the U.S. National Oceanic and Atmospheric Administration (NOAA), has developed a statistical model that simulates the daily maximum and minimum temperatures from three properties of a plot of temperature against time. These three properties are the mean, its daily variance and its day-to-day correlation (the correlation is an indication of how temperatures persist—for

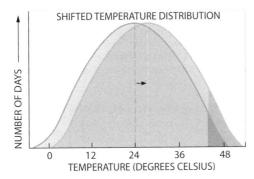

SHIFTED TEMPERATURE DISTRIBUTION

NUMBER OF DAYS

0   12   24   36   48
TEMPERATURE (DEGREES CELSIUS)

Small shifts in the most common daily temperature cause disproportionate increases in the number of extremely hot days. The reason is that temperature distributions are roughly Gaussian. So when the highest point in the Gaussian "bell" curve moves to the right, the result is a relatively large increase in the probability of exceeding extremely high temperature thresholds. A greater probability of high temperature increases the likelihood of heat waves.

example, how often a hot day is followed by another hot day). Given new values of mean, variance and persistence, the model will project the duration and severity of extremes of temperature.

Some of its predictions are surprising. For example, Chicago exhibits considerable variability of temperature from week to week. Even if the mean January temperature went up by four degrees C (an occurrence that may actually take place late in the next century) while the other two properties remained constant, days with minimum temperatures less than −17.8 degrees C (zero degrees Fahrenheit) would still occur. They might even persist for several days in a row. There should also be a significant reduction in the number of early- and late-season freezes. And, not surprisingly, during the summer, uncomfortably hot spells, including so-called killer heat waves, would become more frequent. With just a three degree C increase in the average July temperature, the probability that the heat index (a measure that includes humidity and measures overall discomfort) will exceed 49 degrees C (120 degrees F) sometime during the month increases from one in 20 to one in four.

Because of their effects on agriculture, increases in the minimum are quite significant. Observations over land areas during the latter half of this century indicate that the minimum temperature has increased at a rate more than 50 percent greater than that of the maximum. This increase has lengthened the frost-free season in many parts of the U.S.; in the Northeast, for example, the frost-free season now begins an average of 11 days earlier than it did during the 1950s. A longer frost-free season can be beneficial for many crops grown in places where frost is not very common, but it also affects the growth and development of perennial plants and pests.

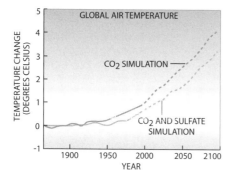

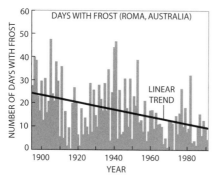

Global air temperature rise was simulated (*above, left*) by a climate model at the U.K. Meteorological Office's Hadley Center. The black line is from a simulation based on carbon dioxide only; the gray line also takes into account sulfate. As the global temperature has increased, the number of days with minimums below zero degrees Celsius has gone down. This example (*below*) shows the annual number of days with frost in Roma, Queensland, in Australia.

The reasons minimum temperatures are going up so much more rapidly than maximums remain somewhat elusive. One possible explanation revolves around cloud cover and evaporative cooling, which have increased in many areas. Clouds tend to keep the days cooler by reflecting sunlight and the nights warmer by inhibiting loss of heat from the surface. Greater amounts of moisture in the soil from additional precipitation and cloudiness inhibit daytime temperature increases because part of the solar energy goes into evaporating this moisture. More conclusive answers, as well as a prediction about whether the asymmetry in daytime and nighttime warming will continue, await better computer models.

Projections of the day-to-day changes in temperature are less certain than those of the mean, but observations have suggested that this variability in much of the Northern Hemisphere's midlatitudes has decreased as the climate has become warmer. Some computer models also project decreases in variability. The variability depends on season and location and is also tied to surface characteristics, such as snow on the ground or moisture in the soil. In midlatitudes, changes in the daily variability of temperature have also been linked to changes in the frequency and intensity of storms and in the location of the paths commonly taken by storms. These storm tracks are, in effect, a succession of eastward-moving midlatitude depressions whose passage dominates the weather.

The relation between these storms and temperature is complex. In a warmer world, the difference of temperature between the tropics and the poles would most likely cover a smaller range, because greater warming is expected near the poles. This factor would tend to weaken storms. On the other hand, high in the atmosphere this difference would be reversed, having the opposite influence. Changes in storms could also happen if anthropogenic aerosols continue to cool the surface regionally, altering the horizontal temperature contrasts that control the location of the storm tracks.

MORE PRECIPITATION

The relation between storms and temperature patterns is one of the reasons it is so difficult to simulate climate changes. The major aspects of climate—temperature, precipitation and storms—are so interrelated that it is impossible to understand one independently of the others. In the global climate system, for example, the familiar cycle of evaporation and precipitation transfers not only water from one place to another but also heat. The heat used at the surface by evaporation of the water is released high in the atmosphere when the water condenses again into clouds and precipitation, warming the surrounding air. The atmosphere then loses this heat by radiating it out into space.

With or without additional greenhouse gases, the earth takes in the same amount of solar energy and radiates the same amount back out into space. With a greater concentration of greenhouse gases, however, the surface is better insulated and can radiate less heat *directly* from the ground to space. The efficiency with which the planet radiates heat to space goes down, which means that the temperature must go up in order for the same amount of heat to be radiated. And as the temperature increases, more evaporation takes place, leading to more precipitation, averaged across the globe.

Precipitation will not increase everywhere and throughout the year, however. (In contrast, all areas of the globe should have warmer temperatures by the end of the next century.) The distribution of precipitation is determined not only by local processes but also by the rates of evaporation and the atmospheric circulations that transport moisture.

For instance, most models predict reduced precipitation in southern Europe in summer as a result of increased greenhouse gases. A significant part of the rainfall in this region comes from local evaporation, with the water not precipitated locally being exported to other areas. Thus, in a warmer climate, increased evaporation in the spring would dry out the soil and lead to less water being available for evaporation and rainfall in the summer.

On a larger scale, most models predict an increase in average precipitation in winter at high latitudes because of greater poleward transport of moisture derived from increased evaporation at low latitudes. Since the turn of the century, precipitation has indeed increased in the high latitudes of the Northern Hemisphere, primarily during the cold season, as temperatures have increased. But for tropical and subtropical land areas, precipitation has actually decreased over the past few decades. This is especially apparent over the Sahel and eastward to Indonesia.

In northernmost North America (north of 55 degrees) and Eurasia, where conditions are normally far below freezing for much of the year, the amount of snowfall has increased over the past several decades. Further increases in snowfall are likely in these areas. Farther south, in southern Canada and the northern U.S., the ratio of snow to rain has decreased, but because of the increase in total precipitation there has been little overall change in the amount of snowfall. In the snow transition belts, where snow is intermittent throughout the cold season, the average snowfall will tend to diminish as the climate warms, before vanishing altogether in some places. Interestingly, areal snow cover during spring and summer abruptly diminished by nearly 10 percent after 1986. This decrease in snow cover has contributed to the rise of spring temperatures in the middle and high latitudes.

Besides the overall amounts of precipitation, scientists are particularly interested in the frequency of heavy downpours or rapid accumulations because of the major practical implications. Intense precipitation can result in flooding, soil erosion and even loss of life. What change do we expect in this frequency?

Whether precipitation occurs is largely determined by the relative humidity, which is the ratio of the concentration of water vapor to its maximum saturation value. When the relative humidity reaches 100 percent, water condenses into clouds, making precipitation possible. Computer models suggest that the distribution of relative humidity will not change much as the climate changes.

The concentration of water vapor needed to reach saturation in the air rises rapidly with temperature, however, at about 6 percent per degree Celsius. So in a warmer climate, the frequency of precipitation (which is related to how often the relative humidity reaches 100 percent) will change less than the amount of precipitation (related to how much water vapor there is in the air). In addition, not only will a warmer world be likely to have more precipitation, but the average precipitation event is likely to be heavier.

Various analyses already support the notion of increased intensity. In the U.S., for example, an average of about 10 percent of the total annual precipitation that falls does so during very heavy downpours in which at least 50 millimeters falls in a single day. This proportion was less than 8 percent at the beginning of this century.

As incredible as it may seem with all this precipitation, the soil in North America, southern Europe and in several other places is actually expected to become drier in the coming decades. Dry soil is of particular concern because of its far-reaching effects, for instance, on crop yields, groundwater resources, lake and river ecosystems and even on down to the foundations of buildings. Higher temperatures dry the soil by boosting the rates of evaporation and transpiration through plants. Several models now project significant increases in the severity of drought. Tempering these predictions, however, are studies of drought frequency and intensity during this century, which suggest that at least during the early stages of global warming other factors have overwhelmed the drying effects of warmer weather. For example, in the U.S. and the former U.S.S.R., increases in cloud cover during the past several decades have led to reduced evaporation. In western Russia, in fact, soil moisture has increased.

STORMY WEATHER

Great as they are, the costs of droughts and heat waves are less obvious than those of another kind of weather extreme: tropical cyclones. These storms, known as hurricanes in the Atlantic and as typhoons in the western North Pacific, can do enormous damage to coastal areas and tropical islands. As the climate warms, scientists anticipate changes in tropical cyclone activity that would vary by region. Not all the consequences would be negative; in some rather arid regions the contribution of tropical cyclones to rainfall is crucial. In northwest Australia, for example, 20 to 50 percent of the annual rainfall is associated with tropical cyclones. Yet the damage done by a single powerful cyclone can be truly spectacular. In August 1992 Hurricane Andrew killed 54 people, left 250,000 homeless and caused $30-billion worth of damage in the Caribbean and in the southeast coastal U.S.

Early discussions of the possible impacts of an enhanced greenhouse effect often suggested more frequent and more intense tropical cyclones. Because these storms depend on a warm surface with unlimited moisture supply, they form only over oceans with a surface temperature of at least 26 degrees C. Therefore, the reasoning goes, global warming will lead to increased ocean temperatures and, presumably, more tropical cyclones.

Yet recent work with climate models and historical data suggests that this scenario is overly simplistic. Other factors, such as atmospheric buoyancy, instabilities in the wind flow, and the differences in wind speed at various heights (vertical wind shear), also play a role in the storms' development. Beyond enabling this rather broad insight, though, climate models have proved of limited use in predicting changes in cyclone activity. Part of the problem is that the simulations are not yet detailed enough to model the very intense inner core of a cyclone.

The historical data are only slightly more useful because they, too, are imperfect. It has been impossible to establish a reliable global record of variability of tropical cyclones through the 20th century because of changes in observing systems (such as the introduction of satellites in the late 1960s) and population changes in tropical areas. Nevertheless, there are good records of cyclone activity in the North Atlantic, where weather aircraft have reconnoitered since the 1940s. Christopher W. Landsea of the NOAA Atlantic Oceanographic and Meteorological Laboratory has documented a decrease in the intensity of hurricanes, and the total number of hurricanes has also followed suit. The years 1991 through 1994 were extremely quiet in terms of the frequency of storms, hurricanes and strong hurricanes; even the unusually intense 1995 season was not enough to reverse this downward trend. It should be noted, too, that the number of typhoons in the northwestern Pacific appears to have gone up.

Overall, it seems unlikely that tropical cyclones will increase significantly on a global scale. In some regions, activity may escalate; in others, it will lessen. And these changes will take place against a backdrop of large, natural variations from year to year and decade to decade.

Midlatitude cyclones accompanied by heavy rainfall, known as extratropical storms, generally extend over a larger area than tropical cyclones and so are more readily modeled. A few studies have been done. A recent one by Ruth Carnell and her colleagues at the Hadley Center of the U.K. Meteorological Office found fewer but more intense storms in the North Atlantic under enhanced greenhouse conditions. But the models do not all agree.

Analyses of historical data also do not give a clear conclusion. Some studies suggest that since the late 1980s, North Atlantic winter storm activity has been more extreme than it ever was in the previous century. Over the past few decades, there has also been a trend toward increasing winds and wave heights in the northern half of the North Atlantic Ocean. Other analyses by Hans von Storch and his colleagues at the Max Planck Institute for Meteorology in Hamburg, Germany, found no evidence of changes in storm numbers in the North Sea. In general, as with the tropical cyclones, the available information suggests that there is little cause to anticipate global increases in extratropical storms but that regional changes cannot be ruled out.

THE FUTURE

Although these kinds of gaps mean that our understanding of the climate system is incomplete, the balance of evidence suggests that human activities have already had a discernible influence on global climate. In the future, to reduce the uncertainty regarding anthropogenic climate change, especially on the small scales, it will be necessary to improve our computer modeling capabilities, while continuing to make detailed climatic observations.

New initiatives, such as the Global Climate Observing System, and detailed studies of various important climatic processes will help, as will increasingly powerful supercomputers. But the climate system is complex, and the chance always remains that surprises will come about. North Atlantic currents could suddenly change, for example, causing fairly rapid climate change in Europe and eastern North America.

Among the factors affecting our predictions of anthropogenic climate change, and one of our greatest uncertainties, is the amount of future global emissions of greenhouse gases, aerosols and other relevant agents. Determining these emissions is much more than a task for scientists: it is a matter of choice for humankind.

FURTHER READING

CHANGING BY DEGREES: STEPS TO REDUCE GREENHOUSE GASES. U.S. Congress, Office of Technology Assessment, 1991.

POLICY IMPLEMENTATION OF GREENHOUSE WARMING: MITIGATION, ADAPTATION, AND THE SCIENCE BASE. National Academy of Sciences. National Academy Press, 1992.

GLOBAL WARMING DEBATE. Special issue of *Research and Exploration: A Scholarly Publication* of the National Geographic Society, Vol. 9, No. 2; Spring 1993.

GLOBAL WARMING: THE COMPLETE BRIEFING. John T. Houghton. Lion Press, 1994.

CLIMATE CHANGE 1995: THE SCIENCE OF CLIMATE CHANGE. Contribution of Working Group I to the *Second Assessment Report of the Intergovernmental Panel on Climate Change.* Edited by J. T. Houghton, L. G. Meira Filho, B. A. Callendar and N. Harris. Cambridge University Press, 1996.

INDICES OF CLIMATE CHANGE FOR THE UNITED STATES. T. R. Karl, R. W. Knight, D. R. Easterling and R. G. Quayle in *Bulletin of the American Meteorological Society,* Vol. 77, No. 2, pages 279–292; February 1996.

Tsunami: Wave of Change

ERIC L. GEIST, VASILY V. TITOV AND COSTAS E. SYNOLAKIS

ORIGINALLY PUBLISHED IN JANUARY 2006

O n December 26, 2004, a series of devastating waves attacked coastlines all around the Indian Ocean, taking the largest toll of any tsunami ever recorded. The surges decimated entire cities and villages, killing more than 225,000 people within a matter of hours and leaving at least a million homeless.

This shocking disaster underscored an important fact: as populations boom in coastal regions worldwide, tsunamis pose a greater risk than ever before. At the same time, this tsunami was the best documented in history—opening a unique opportunity to learn how to avoid such catastrophes in the future. From home videos of muddy water engulfing seaside hotels to satellite measurements of the waves propagating across the open ocean, the massive influx of information has reshaped what scientists know in several ways.

For one thing, the surprising origin of the tsunami—which issued from a location previously thought unlikely to birth the giant waves—has convinced researchers to broaden their list of possible danger areas. The new observations also provided the first thorough testing of computer simulations that forecast where and when a tsunami will strike and how it will behave onshore. What is more, this event revealed that subtle complexities of an earthquake exert a remarkably strong influence over a tsunami's size and shape. The improved models that have resulted from these discoveries will work with new monitoring and warning systems to help save lives.

BEFORE THE BIG ONE

Researchers have long known that the breeding grounds for nearly all tsunami-generating earthquakes are subduction zones. Marked by immense trenches on the seafloor, such areas form where one of the planet's outer shell of tectonic plates plunges underneath another. Gravitational forces and the motion of viscous material deep within the earth's mantle work to keep the plates moving past each other, but friction in the shallow crust temporarily locks them together. As a result, stress builds up across the vast interface, or fault, between the two plates. Sometimes that stress is relieved suddenly in the form of a large earthquake. The bottom plate dives farther down, snapping the top plate violently upward—and the overlying seawater goes along for the ride. The size of the resulting tsunami depends on how much the seafloor moves. Once generated, the tsunami splits in two; one moves quickly inland while the second heads toward the open ocean.

In the eastern Indian Ocean, off the west coast of Sumatra, Indonesia, the India Plate slips below the Eurasia Plate along the Sumatra subduction zone. Southern parts of this fault zone have produced large (magnitude 9) earthquakes in the past, most recently in 1833; Kerry Sieh of the California Institute of Technology and his colleagues have found ancient coral reefs uplifted by these events. Experts were on the watch for another large shock there.

But they were puzzled when the tsunami-producing event of December 2004 originated in the upper part of this region, just northwest of Sumatra. Prior records had shown much slower motion along the offshore fault there, so it was unclear that stress could ever build up enough to result in such a violent tremor. Yet later analysis revealed that the magnitude 9 shock raised a 1,200-kilometer stretch of seafloor by as much as eight meters in some places as it unlocked a California-size area of the fault zone—displacing hundreds of cubic kilometers of seawater above normal sea level. As a result, investigators now are considering possible additional tsunami threats near Alaska, Puerto Rico and other similar subduction zones.

The Sumatra-Andaman shock began at 7:59 A.M. local time, and soon a global network of seismic stations alerted the Pacific Tsunami Warning Center in Ewa Beach, Hawaii. Although geophysicists there were some of the first people outside the region to learn of the earthquake, they had no way to confirm that a deadly tsunami was surging across the Indian Ocean until they got the first news bulletin of the catastrophe unfolding.

In the Pacific Ocean, where 85 percent of the world's tsunamis occur, remote sensors called tsunameters can detect a tsunami offshore and warn the Pacific center's scientists and those at a second center in Palmer, Alaska, before the waves make landfall [see "Tsunami!" by Frank González, pp. 195–206]. But no such technology was in place for the Indian Ocean, and no established lines of communication existed to transmit a warning to people on the coast. Although the first waves took two hours or more to reach Thailand, Sri Lanka and many of the areas hardest hit, nearly everyone was taken by surprise.

IN THE OPEN OCEAN

That December day forever changed the world's appreciation for how much damage tsunamis can inflict, where they might strike and how utterly defenseless so many communities are. International groups have been scrambling to rectify the situation ever since. Meanwhile researchers have been digging through the clues this disaster left behind to sharpen their understanding of how tsunamis get started, propagate and then crash against the shores—and to better warn of the next one.

For 15 years, researchers in Japan and the U.S. have been developing computer models that simulate how a tsunami propagates through the open ocean. Before now, however, investigators had few observations to compare against their theories. All tsunami-propagation models require two key starting variables: an estimate of the location and area of deformed seafloor, which researchers base on the quake's magnitude and epicenter, and a measure of the height, or amplitude, of the displaced water. The latter can be inferred adequately for real-time forecasts only after making direct observations of tsunami waves in the open ocean.

But for previous major tsunamis, scientists just had measurements that tide gauges recorded near the coast or that surveyors later estimated from water damage on land. The main problem is that near shore, a tsunami's actual size is masked by additional waves generated as the tsunami bounces off seawalls, wraps around islands, or sloshes back and forth in a bay—all of which make for a very tangled signal.

By sheer coincidence, a trio of earth-monitoring satellites gave modelers the pristine, undistorted wave heights they needed for the Indian Ocean tsunami. The satellites happened to orbit over the region between two and nine hours after the earthquake, making the first radar measurements of a tsunami propagating across the open ocean. The results proved for the first time that—as suspected—a bump of water only half a meter high in the open ocean can truly transform into the towering surges that wreak so much destruction on land.

With a ground speed of about 5.8 kilometers a second, the satellites also provided the first continuous transect of tsunami amplitudes—that is, they monitored the waves continuously along their travel path rather than making measurements in a single spot, as tide gauges do. As it turns out, the modeled and measured wave heights match each other quite well, validating general theories about how tsunamis move across the open ocean—and confirming that the current modeling paradigms are a useful tool for public safety, even for the largest tsunami.

GLOBAL REACH

The global scope of the tsunami further corroborated that the models are sound for forecasting. Because a tsunami in the deep ocean moves along at about the same speed as a jet airliner (from 500 to 1,000 kilometers an hour), the first wave took less than three hours to travel from northern Sumatra and the Andaman Islands east to Myanmar (Burma), Thailand and Malaysia and west to Sri Lanka, India and the Maldives. Within 11 hours it had struck the South African coast 8,000 kilometers away, the farthest point away where a tsunami-related death was reported.

But the waves did not stop there. About the same time the tragedy made the news, scientists started getting records from tide gauge stations around the world. On its westward path, the tsunami curved around the southern tip of Africa and then split as it traveled northward through the Atlantic; one branch headed toward Brazil and the other toward Nova Scotia. On its eastward path, the tsunami sped through the gap between Australia and Antarctica and into the Pacific as far north as Canada. Not since the eruption of Krakatau volcano in 1883 had a tsunami been known to travel so far.

When the entire path of the tsunami played out on the National Oceanic and Atmospheric Administration's leading computer simulation, called MOST (short for *method of splitting tsunami*), the simulated wave heights agreed quite well with the measurements at various tide gauge stations. What is more, the model revealed just *how* a tsunami manages to travel so far. A map of the simulated wave heights for the Indian Ocean event showed that they are highest along mid-ocean ridges. These ridges, which connect one ocean basin to the next, appear to channel the wave energy farther than it might otherwise travel. Knowing about this effect is helpful for forecasting because modelers can better estimate where the strongest wave energy is most likely to go.

IMMEDIATE AFTERMATH

Forecasting how a tsunami will behave once it climbs ashore is a much greater challenge. As always occurs in tsunamis, the December event's waves gradually slowed down as they entered shallow water. By the time the ripples reached shore, the distance between crests, which was hundreds of kilometers in the open ocean, had decreased to 15 or 20 kilometers. But with the fast water still pushing from behind, the wave peaks grew higher and higher, to more than 30 meters in Sumatra's Aceh province, the first region hit.

Still moving at about 30 to 40 kilometers an hour, the waves swept inland—more than four kilometers in parts of the city of Banda Aceh. They receded just as violently, carrying far out to sea anything they picked up on the way in. Along all inundated shorelines, waves pounded the coasts for hours. And with 30 minutes or more between crests, many people haplessly returned to the beaches, only to be attacked by subsequent waves. The cumulative damage to the physical environment was so vast that it was visible to astronauts in space; it was also extremely varied.

Considering the many factors involved, how could models reliably predict such variation? Until the early 1990s, because of unresolved computational complexities, even the best simulations ended their calculations at the water's edge or just offshore. Investigators then used that last height to estimate how far inland a tsunami would climb. But the initial careful surveys of tsunami disasters suggested their estimates were way off. A tsunami that struck Nicaragua in 1992 was the first time scientists made comprehensive field measurements to compare with the model predictions. The flooding levels in some places were up to 10 times higher than what the models had predicted.

A race of sorts soon developed between U.S. and Japanese modelers seeking to describe the inundation more accurately, by calculating the entire evolution of the tsunami on dry land. Through a combination of large-scale laboratory experiments and field measurements from subsequent tsunamis, investigators refined the Japanese TUNAMI-N2 and the U.S. MOST models until they could match the inundation patterns of most past tsunamis quite well—as long as high-resolution data about the coastal and offshore topography were available. These researchers did not know, however, that the models would work as well for the largest tsunamis. As it turned out, the models matched the Indian Ocean flooding better than expected, despite the relative lack of detail about the coastal landscape.

Post-tsunami surveyors in Indonesia and elsewhere quickly noted that predictions of floodwater depth alone could not always foretell the full impact of the tsunami. In many locales in Thailand and Sri Lanka, the tsunami depth on land was less than 4.5 meters, yet the devastation rivaled that in Aceh, where the water was six times deeper. Another shocking reality was that in Banda Aceh, the waves destroyed block after block of reinforced concrete structures that may have withstood the earthquake's shaking.

To account for the magnitude of the wreckage, Ahmet C. Yalciner of Middle East Technical University in Ankara, Turkey, and one of us (Synolakis) are devising new damage metrics—standards that coastal engineers can use to assess the force of tsunami waves on structures—that also consider powerful currents, which are much stronger in tsunami floodwaters than they are in normal tides and storm waves.

SHAKING SURPRISES

Arguably the greatest scientific conundrum regarding the Indian Ocean tsunami is the earthquake itself. Even the magnitude of the earthquake is still debated, with some estimates as high as 9.3. Although the seismic shock was the largest since the 1964 Alaska earthquake, it has been a challenge to describe how the Sumatra-Andaman fault produced such a giant tsunami.

By any measure, this earthquake was amazingly complex. Typically the fault slip will be largest right at the start, near its origin. Yet in some cases, the fault break begins with a small amount of slip, suggesting that the earthquake will be small, then hits a weak or highly stressed part of the fault that lets loose violently, resulting in a much larger earthquake and tsunami. That is what happened in the 2004 tsunami. Such cases are challenging to analyze in time to make a useful warning.

NOAA's tsunami forecast models were put to the test for this perplexing event. Running the model with seismic data alone would have underestimated tsunami heights in the open ocean by a factor of 10 or more. Adding the first direct measurement of tsunami amplitude, which reached scientists from a tide gauge station at Cocos Island about three and a half hours after the earthquake occurred, improved the results dramatically. But something was still missing.

In the days following the earthquake, analyses of the shock's strong seismic waves indicated that the initial fault break sped northward from Sumatra at 2.5 kilometers a second. They also pinpointed the areas of greatest slip—and thus of the greatest tsunami generation. The problem for tsunami modelers was that none of these seismic solutions included enough overall fault motion to reproduce either the satellite observations of wave heights in the open ocean or the severe flooding in Banda Aceh.

The critical clue came from land-based stations that use the Global Positioning System (GPS) to track ground movements much slower than what seismic waves produce. Those measurements revealed that the fault continued to slip, albeit slowly, after it stopped emanating seismic energy. Although there is a limit to how slowly a fault can slip and still generate a tsunami, it is most likely that this often overlooked phenomenon, called after-slip, accounts for the surprising tsunami heights. If so, incorporating continuous GPS readings may be an important component of tsunami warning systems in the future.

HIT OR MISS

The specific factors in any given earthquake clearly exert fearsome controls on tsunamis. As if emphasizing this point, Planet Earth produced another huge temblor along the same fault on March 28, 2005. The initial break occurred an equivalent distance from the Sumatra coastline and at virtually the same depth below the seafloor as the December earthquake, and both shocks are among the top 10 largest recorded since 1900. Yet they produced radically different tsunamis.

Seeing the March earthquake flash on their computer screens as a magnitude 8.7, scientists at the Pacific Tsunami Warning Center and elsewhere expected the worst. Severe damage from strong ground shaking indeed occurred but without immediate reports of tsunami damage. When an international team (including one of us, Titov) surveyed the region two weeks later, they measured tsunami run-up heights as high as four meters—still potentially deadly. Some Indonesians said they learned from their first experience and ran inland when the ground shook. Better evacuation is only one reason the March tsunami did not take more lives.

Analysis of aftershocks from the December earthquake suggested to Andrew Newman of the Georgia Institute of Technology and Susan Bilek of the New Mexico Institute of Mining and Technology that the fault slipped near the deep trench that time and was thus under deeper water than the main part of the fault that slipped in March was. The December tsunami thus had more opportunity to gain height during its trip from deep water to shore. In addition, unlike the December tsunami, fault movement in March occurred below the islands of Nias and Simeulue, thereby limiting the amount of water the uplifting crust could displace.

A slight difference in the fault orientation meant that their tsunami waves proceeded in two different general directions. For the March earthquake, most of its eastbound waves smashed into the island of Sumatra, which blocked much of the wave energy from moving on toward Thailand and Malaysia. The westbound waves shot out into the open ocean to the southwest, largely missing Sri Lanka, India and the Maldives, all of which suffered terribly in December. These examples highlight the critical importance that even small variations in the location of an earthquake can make.

Despite the lingering scientific uncertainties that will probably always surround such complex phenomena, the new tsunami science is ready for implementation. The biggest challenge for saving lives is now applying the scientific findings to proper education, planning and warning.

FURTHER READING

FURIOUS EARTH: THE SCIENCE AND NATURE OF EARTHQUAKES, VOLCANOES, AND TSUNAMIS. Ellen J. Prager. McGraw-Hill, 2000.

National Oceanic and Atmospheric Administration tsunami pages: www.tsunami.noaa .gov/

University of Southern California Tsunami Research Center: http://cwis.usc.edu/dept/ tsunamis/2005/index.php

U.S. Geological Survey Tsunami and Earthquake Research: http://walrus.wr.usgs.gov/ tsunami/

THE OCEANS IN PERIL

Enriching the Sea to Death

SCOTT W. NIXON

ORIGINALLY PUBLISHED IN *SCIENTIFIC AMERICAN PRESENTS*, FALL 1998

The widespread pollution of Narragansett Bay began with a great celebration on Thanksgiving Day, 1871. For 10 full minutes, the church
bells of Providence, R.I., rang out, and a 13-gun salute sounded. The townspeople were giving thanks for the completed construction of their first
public water supply. Soon afterward clean water flowed through taps and
flush toilets, liberating residents forever from backbreaking trips to the
well and freezing visits to the privy. Millions learned the joys of running
water between about 1850 and 1920, as towns throughout North America
and Europe threw similar parties. But homeowners gave scant thought to
how their gleaming new water closets would change the makeup of the
oceans.

With the wonder of running water came the unpleasant problem of
running waste. No longer was human excrement deposited discreetly in
dry ground; the new flush toilets discharged streams of polluted water
that often flowed through the streets. Town elders coped with the un-
happy turn of events by building expensive networks of sewers, which
invariably routed waste to the most convenient body of water nearby. In
this way, towns quickly succeeded in diverting the torrent of waste from
backyards and city streets to fishing spots, swimming holes and adjacent
ocean shores. In many cases, the results were disastrous for the aquatic
environment. And as the flow continues, society still struggles with the
repercussions for the plants and animals that inhabit coastal waters.

UNTAMED GROWTH

Even a century ago the unsightly consequences of dumping raw sewage directly into lakes and bays were quite troubling. Dead fish and malodorous
sludges fouled favorite beaches as sewage rode back toward land on the
waves. Unwilling to return to the days of chamber pots and privies, people
were soon forced to clean up their waste somewhat before discharging it.

The wastewater-treatment technologies put into place between about 1880 and 1940 removed visible debris and pathogenic organisms from sewer effluent, effectively eliminating the distasteful reminders that had once washed up on the shore. By the 1960s many treatment plants had begun to remove organic matter as well. But the various methods failed to extract the elements nitrogen and phosphorus, nutrients indispensable to human life and abundant in human waste. These invisible pollutants were flushed into rivers, lakes and oceans in prodigious quantities, and no telltale sign heralded the harm they could inflict.

As every farmer and gardener knows, nitrogen and phosphorus are the essential ingredients of plant fertilizers. Plants that live underwater often respond to these nutrients just as beets and roses do: they grow faster. Of course, aquatic plants are different from the trees and shrubs familiar to landlubbers—most are microscopic, single-celled organisms called phytoplankton that drift suspended in the currents.

Where nutrients are scarce, phytoplankton are sparse and the water is usually crystal-clear. But in response to fertilization, phytoplankton multiply explosively, coloring the water shades of green, brown and red with their photosynthetic pigments. These blooms increase the supply of organic matter to aquatic ecosystems, a process known as eutrophication.

Pollution-driven eutrophication was not recognized as a serious threat to many larger lakes in Europe and North America until the 1950s and 1960s—Lakes Erie and Washington in the U.S. are well-known examples. Why was the accelerating growth of phytoplankton a concern? After all, people welcomed the "green revolution" that fertilizers helped to bring to agriculture around that time. The difference underwater results from the precarious balance between oxygen supply and demand in aquatic ecosystems.

Terrestrial ecologists do not usually worry about oxygen, because the air is full of it: each cubic meter contains some 270 grams. And the atmosphere is constantly in motion, replenishing oxygen wherever it is used. But water circulates less readily than air and holds only five to 10 grams of oxygen per cubic meter at best—that is, when freely exchanging its dissolved gases with the atmosphere. Although fish and a number of other aquatic animals have adapted to live under these conditions, a small decrease in the oxygen content of their surroundings can often be deadly to them.

Phytoplankton floating near the surface of nutrient-rich lakes fare better in the oxygen equation. They receive ample sunlight to carry out photosynthesis during the day and have access to plenty of oxygen to support

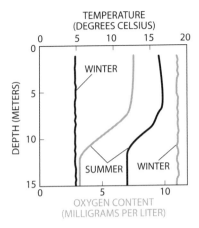

TEMPERATURE (DEGREES CELSIUS)

DEPTH (METERS)

OXYGEN CONTENT (MILLIGRAMS PER LITER)

Sea grass is smothered by the macroalgae growing in shallow Danish waters enriched with nutrient runoff from farms and sewers. In winter, the water is well mixed, and oxygen levels are uniform throughout (*left*). In summer, warm, sluggish surface water floats on top of cooler, saltier bottom water, which rapidly becomes depleted of oxygen as phytoplankton decay.

their metabolism at night. But even under the best circumstances, phytoplankton are short-lived: the tiny organisms continually die off and sink, leaving new generations growing in their place. The more abundant the bloom, the heavier the fallout to the lower depths. And therein lies the problem: the bottom-living bacteria that digest this dead plant matter consume oxygen.

When organic material is abundant in a lake and where surface and bottom waters seldom mix—for example, where winds are calm—oxygen rapidly becomes scarce below the surface. Animals that cannot escape to better-aerated zones will suffocate, and dead creatures may begin to litter the shoreline as bacteria take over the otherwise barren bottom waters. During the 1970s, such awful conditions used to regularly overcome oxygen-starved Lake Erie, which was said to be "dying."

DEAD ZONES

Until about 40 years ago, the oceans were thought to be immune to the combined forces of nutrient enrichment and oxygen depletion, which were then commonly observed at work in lakes. After all, the seas are vast and restless—the waste discharged from land seemed just a drop in a giant, sloshing bucket.

Scientists now know this assumption was wrong. The fertilization of coastal waters constitutes a major environmental threat to the Baltic Sea, the Gulf of Mexico, Chesapeake Bay, the Lagoon of Venice, the North Sea and a great many other estuaries, bays and lagoons in the industrial world. Most at risk are sheltered regions that do not experience winds or tides strong enough to keep the sea thoroughly mixed the whole year around. For just like nutrient-rich lakes, polluted bays and estuaries can become starved of oxygen when their bottom waters are cut off from the atmosphere.

Coastal areas are especially vulnerable to oxygen depletion because freshwater draining into the ocean from rivers and streams—often laden with nutrients—tends to float on top of denser saltwater. In summer, the surface layer becomes even more buoyant as it warms in the sun. Unless some energetic mixing ensues, the lighter, oxygen-rich veneer will remain isolated from the denser water below. In areas of weak wind and tide, such stratification can last an entire summer.

When a polluted bay or estuary remains relatively still for weeks, months or whole seasons, the difference between life at the top and life at the bottom becomes stark. The surface waters, rich in nutrients and bathed in sunlight, teem with phytoplankton and other forms of floating plant life. The bottom layers become choked with dead plant matter, which consumes more and more oxygen as it decomposes. Below the surface, entire bays can suffocate. And the problem is not necessarily limited to protected waters near the shore. Where enough nutrients arrive and currents are configured just right, even open waters can fall victim. For instance, oxygen deprivation cuts a lethal swath through some 18,000 square kilometers (7,000 square miles) of the deep waters of the Gulf of Mexico every summer, creating a barren region called the "dead zone."

The effects of eutrophication trickle up into human affairs in various ways. Bays and estuaries provide some of the richest fishing grounds, yet oxygen depletion kills fish, and nutrients may cause certain toxic varieties of phytoplankton to bloom, contaminating the shellfish that feed on them. Picturesque shores are sullied by dead fish and rotting plant waste, and the water may reek of rotten eggs as bacteria on the ocean floor spew out hydrogen sulfide.

Fertilization of coastal waters also changes life underwater in more subtle ways. For example, as the balance of nutrients changes, the mix of phytoplankton may shift in response. In particular, diatoms, which need about as much silicon as nitrogen, cannot benefit. Because pollution increases the supply of nitrogen but not the amount of silicon, these important organisms may be crowded out by other species of phytoplankton that are less useful to feeding fish and shellfish.

What is more, sunlight does not penetrate deeply into water clouded by blooms. Thick layers of phytoplankton may shade out the sea grasses and seaweeds that typically grow in coastal waters and shelter vulnerable creatures such as crabs and young fish. As a result, complex aquatic food chains may be broken apart.

CATTLE, CORN AND CARS

The assault on the waters of the developed world that began with urban sewage systems in the mid-1800s has only escalated since that time. Because nitrogen and phosphorus are essential for human nutrition, the rapidly growing world population consumes—and excretes—ever larger amounts of both elements. This factor alone almost doubled the release of nutrients from human waste between 1950 and 1985. And not only are there more people on the earth but also the typical diet is becoming ever richer in protein. All this protein contains abundant nitrogen, which just increases the burden on the environment when it is metabolized and finally excreted.

As the human population has skyrocketed, so has the number of animals raised for food. The count of livestock—animals that also consume and excrete large amounts of nitrogen and phosphorus—has grown by 18 percent during the past 20 years. To produce the huge quantities of crops needed to feed both humans and livestock, farmers have been applying exponentially increasing amounts of fertilizer to their fields since the 1950s. The main ingredients in these fertilizers are nitrogen and phosphorus. Rain washes these nutrients off the land and into rivers and streams, which then carry them to lakes and oceans.

Between 1960 and 1980 the application of nitrogen fertilizer increased more than fivefold, and in the decade that followed, more synthetic fertilizer was spread on land than had been applied throughout the entire previous history of agriculture. Farmers have also been raising increasing quantities of legumes (such as soybeans), which live in partnership with microorganisms that convert nitrogen to nutritive forms. Vast quantities of enriching nitrogen compounds—perhaps equal to half of what is produced as fertilizer—have become available from this source.

That there have been widespread changes in the oceans is not surprising. The dead zone that forms in the Gulf of Mexico every summer probably results from excess fertilizer washed from farms and carried down the Mississippi. Unfortunately, such nutrient injections may be even more dangerous to coastal waters than to lakes. Research early on showed that phosphorus rather than nitrogen induces aquatic plants to bloom in most freshwater environments. This news was in a sense good for lakes, because phosphorus is more easily managed than nitrogen.

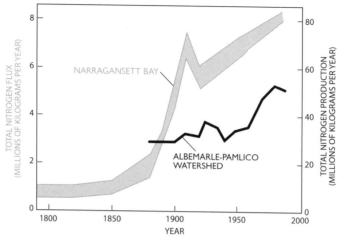

Nitrogen levels in Narragansett Bay, R.I., increased dramatically before the turn of the century, after installation of a public water supply and sewer system. Certain other watersheds did not experience a sharp rise in nitrogen levels until fertilizer use took off in the 1950s and 1960s; an example is the Albemarle-Pamlico watershed in North Carolina.

Phosphorus is chemically sticky and binds easily to other substances. Thus, it tends to adhere to soil and is less likely than nitrogen to leach out of fertilized fields. And phosphorus can be easily removed from sewage by taking advantage of the same stickiness: during treatment, chemicals are added that bind up the element and then settle out along with other sludge. Largely because of improved phosphorus removal from sewage and a widespread ban on the use of phosphate in products such as laundry detergent, the eutrophication of many lakes and rivers has been stopped or greatly reduced.

There is increasing evidence that the phytoplankton of most temperate estuaries, bays and other coastal ocean waters respond not so much to phosphorus as to nitrogen. Marine scientists still do not fully understand the reasons for this difference, but the implications are quite profound. Nitrogen washes easily from fertilized fields into streams and rivers; many sewage-treatment plants are not yet configured to remove nitrogen from wastewater; and there is an additional, copious supply of nitrogen to the oceans—the atmosphere.

Lightning has always converted a tiny amount of inert nitrogen gas, which makes up 78 percent of air, into soluble compounds that plants can take up in their roots and metabolize. But the combustion of fossil fuels has unleashed a torrent of such nitrogen compounds into the atmo-

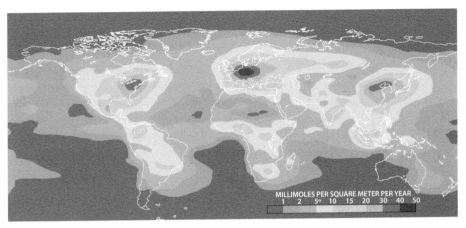

Atmospheric deposition of reactive nitrogen has large geographic variations over the continents and oceans, as shown by computer modeling (*above*). Most of this nutritive nitrogen comes from the combustion of fossil fuels in highly developed areas.

sphere. When oil, gas and coal burn at high temperatures in engines and electric-power generators, they produce nitrogen oxides. Rain and wind carry these soluble compounds to the earth, further enriching coastal waters already replete with sewage and agricultural runoff. In all, fossil-fuel combustion accounts for about 15 percent of the biologically available nitrogen that human activities add to the world every year.

FUTURE SHOCK

In the 1990s marine eutrophication remains a problem of many wealthy nations. Countries such as the U.S. spend billions on fertilizer, automobiles, power plants and sewer systems, all of which feed nitrogen into the oceans. In fact, the amount of nitrogen available per square kilometer of land from fertilizer application, livestock and human waste alone is currently more than 100 times greater in Europe than in much of Africa.

Fortunately, at least the richer nations may be able to afford high-tech remedies. Sewage-treatment facilities that can eliminate nitrogen from wastewater are springing up, and man-made wetlands and precision application of fertilizers may stem the flow from farms. But just as people are seeing improvements in some of the worst-polluted coastal waters in the U.S. and Europe, the developing world is poised to repeat what industrial countries experienced over the past 100 years.

Part of the problem will come directly as a result of population growth. With the occupancy of the planet set to reach more than nine billion by

2050, there will be that many more mouths to feed, more fields to fertilize, more livestock to raise and more tons of waste to dispose of. Many experts predict that the release of nutritive nitrogen from fertilizer and fossil-fuel combustion will double in the next 25 years, most of that increase occurring in the developing world.

The United Nations Population Fund estimates that 80 percent of the rise in global population is taking place in the urban areas of Africa, Asia and Latin America. This increase amounts to about 81 million more people every year, a situation akin to spawning 10 cities the size of Moscow or Delhi. Compounding this source of urban growth is the continuing movement of people from the countryside into cities. It was city sewers that first overloaded waterways such as Narragansett Bay with nutrients, and the scenario is not likely to play out differently in the developing world. Sewers there, too, will likely carry raw sewage initially, and where treatment of these sludges does occur, it will probably not remove nitrogen for many years.

With large stretches of coastline exposed to unprecedented levels of nitrogen, it seems inevitable that ocean waters around the world will become greener, browner and redder and that there will be more frequent periods when the bottom of the sea in vulnerable locations becomes lifeless. Much of the next round of pollution will take place in the waters of the tropics, where both the corals and the fish that inhabit these delicate ecosystems are at risk. Yet it remains difficult to gauge exactly how damaging this inadvertent fertilization will ultimately prove. Scientists are still far from understanding all the ways the oceans will pay for keeping human life so widespread and abundant. But as far as the residents of the ocean are concerned, there seems little cause for celebration.

FURTHER READING

COASTAL MARINE EUTROPHICATION: A DEFINITION, SOCIAL CAUSES, AND FUTURE CONCERNS. Scott W. Nixon in *Ophelia*, Vol. 41, pages 199–219; February 1995.
GLOBAL POPULATION AND THE NITROGEN CYCLE. Vaclav Smil in *Scientific American*, Vol. 277, No. 1, pages 76–81; July 1997.
GLOBAL NITROGEN OVERLOAD PROBLEM GROWS CRITICAL. A. S. Moffat in *Science*, Vol. 279, pages 988–989; February 13, 1998.

Red Tides

DONALD M. ANDERSON

ORIGINALLY PUBLISHED IN AUGUST 1994

Late in 1987 scientists faced a baffling series of marine catastrophes. First, 14 humpback whales died in Cape Cod Bay, Mass., during a five-week period. This die-off, equivalent to 50 years of "natural" mortality, was not a stranding, in which healthy whales beach themselves. Instead the cetaceans died at sea—some rapidly—and then washed ashore. Postmortem examinations showed that the whales had been well immediately before their deaths and that many of them had abundant blubber and fish in their stomachs, evidence of recent feeding. Alarmed and saddened, the public and press blamed pollution or a chemical spill for the mysterious deaths.

Two more mass poisonings occurred that month, but the victims in these new cases were humans. Fishermen and beachgoers along the North Carolina coast started complaining of respiratory problems and eye irritation. Within days, residents and visitors who had eaten local shellfish experienced diarrhea, dizziness and other symptoms suggesting neurotoxic poisoning. The illnesses bewildered epidemiologists and even prompted public conjecture that a nearby sunken submarine was leaking poison gas.

Concurrently, hospitals in Canada began admitting patients suffering from disorientation, vomiting, diarrhea and abdominal cramps. All had eaten mussels from Prince Edward Island. Although Canadian authorities had dealt with shellfish poisoning outbreaks for decades, these symptoms were unfamiliar and disturbing: some patients exhibited permanent short-term memory loss. They could remember addresses but could not recall their most recent meal, for example. The officials quickly restricted the sale and distribution of mussels but eventually reported three deaths and 105 cases of acute poisoning in humans.

We now know that these seemingly unrelated events were all caused, either directly or indirectly, by toxic, single-celled algae called phytoplankton—vast blooms of which are commonly referred to as red tides. Although red tides have been recorded throughout history, the incidents mentioned above were entirely unexpected. As we shall see, they illustrate several major issues that have begun to challenge the scientific and regulatory communities.

Indeed, there is a conviction among many experts that the scale and complexity of this natural phenomenon are expanding. They note that the number of toxic blooms, the economic losses from them, the types of resources affected and the kinds of toxins and toxic species have all increased. Is this expansion real? Is it a global epidemic, as some claim? Is it related to human activities, such as rising coastal pollution? Or is it a result of increased scientific awareness and improved surveillance or analytical capabilities? To address these issues, we must understand the physiological, toxicological and ecological mechanisms underlying the growth and proliferation of red tide algae and the manner in which they cause harm.

Certain blooms of algae are termed red tides when the tiny pigmented plants grow in such abundance that they change the color of the seawater to red, brown or even green. The name is misleading, however, because many toxic events are called red tides even when the waters show no discoloration. Likewise, an accumulation of nontoxic, harmless algae can change the color of ocean water. The picture is even more complicated: some phytoplankton neither discolor the water nor produce toxins but kill marine animals in other ways. Many diverse phenomena thus fall under the "red tide" rubric.

Of the thousands of living phytoplankton species that make up the base of the marine food web, only a few dozen are known to be toxic. Most are dinoflagellates, prymnesiophytes or chloromonads. A bloom develops when these single-celled algae photosynthesize and multiply, converting dissolved nutrients and sunlight into plant biomass. The dominant mode of reproduction is simple asexual fission—one cell grows larger, then divides into two cells, the two split into four, and so on. Barring a shortage of nutrients or light, or heavy grazing by tiny zooplankton that consume the algae, the population's size can increase rapidly. In some cases, a milliliter of seawater can contain tens or hundreds of thousands of algal cells. Spread over large areas, the phenomenon can be both visually spectacular and catastrophic.

Some species switch to sexual reproduction when nutrients are scarce. They form thick-walled, dormant cells, called cysts, that settle on the seafloor and can survive there for years. When favorable growth conditions return, cysts germinate and reinoculate the water with swimming cells that can then bloom. Although not all red tide species form cysts, many do, and this transformation explains important aspects of their ecology and biogeography. The timing and location of a bloom can depend on when the cysts germinate and where they were deposited, respectively. Cyst production facilitates species dispersal as well; blooms carried into new waters by currents or other means can deposit "seed" populations to colonize previously unaffected areas.

A dramatic example of natural dispersal occurred in 1972, when a massive red tide reaching from Maine to Massachusetts followed a September hurricane. The shellfish toxicity detected then for the first time has recurred in that region virtually every year now for two decades. The cyst stage has provided a very effective strategy for the survival and dispersal of many other red tide species as well.

How do algal blooms cause harm? One of the most serious impacts on human life occurs when clams, mussels, oysters or scallops ingest the algae as food and retain the toxins in their tissues. Typically the shellfish themselves are only marginally affected, but a single clam can sometimes accumulate enough toxin to kill a human being. These shellfish poisoning syndromes have been described as paralytic, diarrhetic and neurotoxic, shortened to PSP, DSP and NSP. The 1987 Canadian outbreak in which some patients suffered memory loss was appropriately characterized as amnesic shellfish poisoning, or ASP. The North Carolina episode was NSP.

A related problem, ciguatera fish poisoning, or CFP, causes more human illness than any other kind of toxicity originating in seafood. It occurs predominantly in tropical and subtropical islands, where from 10,000 to 50,000 individuals may be affected annually. Dinoflagellates that live attached to seaweeds produce the ciguatera toxins. Herbivorous fishes eat the seaweeds and the attached dinoflagellates as well. Because ciguatera toxin is soluble in fat, it is stored in the fishes' tissues and travels through the food web to carnivores. The most dangerous fish to eat are thus the largest and oldest, often considered the most desirable as well.

Symptoms do vary among the different syndromes but are generally neurological or gastrointestinal, or both. DSP causes diarrhea, nausea and vomiting, whereas PSP symptoms include tingling and numbness of the mouth, lips and fingers, accompanied by general muscular weakness. Acute doses inhibit respiration, and death results from respiratory paralysis. NSP triggers diarrhea, vomiting and abdominal pain, followed by muscular aches, dizziness, anxiety, sweating and peripheral tingling. Ciguatera induces an intoxication syndrome nearly identical to NSP.

Illnesses and deaths from algal-derived shellfish poisons vary in number from year to year and from country to country. Environmental fluctuations profoundly influence the growth and accumulation of algae and thus their toxicity as well. Furthermore, countries differ in their ability to monitor shellfish and detect biotoxins before they reach the market. Developed countries typically operate monitoring programs that permit the timely closure of contaminated resources. Illnesses and deaths are thus rare, unless a new toxin appears (as in the ASP crisis in Canada) or an outbreak occurs in an area with no history of the problem (as in North Carolina). Developing countries, especially those having long coastlines or poor populations who rely primarily on the sea for food, are more likely to incur a higher incidence of sickness and death from algal blooms.

Phytoplankton can also kill marine animals directly. In the Gulf of Mexico, the dinoflagellate *Gymnodinium breve* frequently causes devastating fish kills. As the wild fish swim through *G. breve* blooms, the fragile algae rupture, releasing neurotoxins onto the gills of the fish. Within a short time, the animals asphyxiate. Tons of dead fish sometimes cover the beaches along Florida's Gulf Coast, causing several millions of dollars to be lost in tourism and other recreation-based businesses.

Farmed fish are especially vulnerable because the caged animals cannot avoid the blooms. Each year, farmed salmon, yellowtail and other economically important species fall victim to a variety of algal species. Blooms can wipe out entire fish farms within hours, killing fingerlings and large fish alike. Algal blooms thus pose a large threat to fish farms and their insurance providers. In Norway an extensive program is under way to minimize these impacts. Fish farmers make weekly observations of algal concentrations and water clarity. Other parameters are transmitted to shore from instruments on moored buoys. The Norwegian Ministry of Environment then combines this information with a five-day weather forecast to generate an "algal forecast" for fish farmers and authorities. Fish cages in peril are then towed to clear water.

Unfortunately, not much more can be done. The ways in which algae kill fish are poorly understood. Some phytoplankton species produce polyunsaturated fatty acids and galactolipids that destroy blood cells. Such an effect would explain the ruptured gills, hypoxia and edema in dying fish. Other algal species produce these hemolytic compounds and neurotoxins as well. The combination can significantly reduce a fish's heart rate, resulting in reduced blood flow and a deadly decrease in oxygen.

Moreover, nontoxic phytoplankton can kill fish. The diatom genus *Chaetoceros* has been linked to dying salmon in the Puget Sound area of Washington State, yet no toxin has ever been identified in this group. Instead species such as *C. convolutus* sport long, barbed spines that lodge between gill tissues and trigger the release of massive amounts of mucus. Continuous irritation exhausts the supply of mucus and mucous cells, causing lamellar degeneration and death from reduced oxygen exchange. These barbed spines probably did not evolve specifically to kill fish, since only caged fish succumb to the blooms. The problems faced by fish farmers are more likely the unfortunate result of an evolutionary strategy by certain *Chaetoceros* species to avoid predation or to stay afloat.

Algal toxins also cause mortalities as they move through the marine food web. Some years ago tons of herring died in the Bay of Fundy after consuming small planktonic snails that had eaten the PSP-producing dinoflagellate *Alexandrium*. From the human health standpoint, it is fortunate that herring, cod, salmon and other commercial fish are sensitive to these toxins and, unlike shellfish, die before toxins reach dangerous levels in their flesh. Some toxin, however, accumulates in the liver and other organs of certain fish, and so animals such as other fish, marine mammals and birds that consume whole fish, including the viscera, are at risk.

We now can reconstruct the events that killed the whales in 1987. A few weeks of intense investigations that year by marine pathologist Joseph R. Geraci of the Ontario Veterinary College, myself and many others revealed that the PSP toxins most likely caused these deaths. The dinoflagellate *Alexandrium tamarense* produced the toxins, which reached the whales via their food web. We analyzed mackerel that the whales had been eating and found saxitoxin, not in their flesh but concentrated in the liver and kidney. Presumably the mackerel ate zooplankton and small fish that had previously dined on *Alexandrium*.

The humpbacks were starting their southward migration and were feeding heavily. Assuming that they consumed 4 percent of their body weight daily, we calculated that they received a saxitoxin dosage of 3.2 micrograms per kilogram of body weight. But was this a fatal dose? Unfortunately, in 1987 we had no data that directly addressed how much toxin would kill a whale. We knew the minimum lethal dose of saxitoxin for humans is seven to 16 micrograms per kilogram of body weight, but that was two to five times more than what the whales had probably ingested.

Our calculations were initially disheartening, but as we thought about it we realized that whales might be more sensitive to the toxins than are humans. First, whales would have received continual doses of toxin as they fed, whereas human mortality statistics are based on single feedings. Second, during a dive, the mammalian diving reflex channels blood and oxygen predominantly to the heart and brain. The same mechanism sometimes protects young children who fall through thin ice and survive drowning, despite being underwater for half an hour or longer. For humans, cold water induces the reflex, but for whales, it is activated during every dive.

Each dive then would expose the most sensitive organs to the toxin, which would bypass the liver and kidney, where it could be metabolized and excreted. Finally, saxitoxin need not have killed the whales directly. Even a slightly incapacitated animal might have difficulty orienting to the water surface or breathing correctly. The whales may actually have drowned following a sublethal exposure to saxitoxin. The exact cause will never be known, but the evidence strongly suggests that these magnificent creatures died from a natural toxin originating in microscopic algae.

Other examples of toxins traveling up the food web appear nearly every year. In 1991 sick or dying brown pelicans and cormorants were found near Monterey Bay, Calif. Wildlife experts could find no signs that pesticides, heavy metals or other pollutants were involved. The veterinarian in charge of the study telephoned Jeffrey Wright of the National Research Council laboratory in Halifax, Nova Scotia. Wright had directed the Canadian Mussel Toxin Crisis Team that identified the poison responsible for the mysterious ASP episode in 1987. His team had isolated a toxin from the Prince Edward Island mussels, called domoic acid, and traced it to its source—a diatom, *Pseudonitzschia pungens,* that had been considered harmless. Four years later members of the same Canadian team quickly ascertained that the sick and dying birds in California had eaten anchovies that contained domoic acid, again from *Pseudonitzschia* (but a different species).

The toxins responsible for these syndromes are not single chemical entities but are families of compounds having similar chemical structures and effects. For example, the saxitoxins that cause PSP are a family of at least 18 different compounds with widely differing potencies. Most algal toxins cause human illness by disrupting electrical conduction, uncoupling communication between nerve and muscle, and impeding critical physiological processes. To do so, they bind to specific membrane receptors, leading to changes in the intracellular concentration of ions such as sodium or calcium.

The saxitoxins bind to sodium channels and block the flux of sodium in and out of nerve and muscle cells. Brevetoxins, the family of nine compounds responsible for NSP, bind to a different site on the sodium channel but cause the opposite effect from saxitoxin. Domoic acid disrupts normal neurochemical transmission in the brain. It binds to kainate receptors in the central nervous system, causing a sustained depolarization of the neurons and eventually cell degeneration and death. Memory loss in ASP victims apparently results from lesions in the hippocampus, where kainate receptors abound.

Why do algal species produce toxins? Some argue that toxins evolved as a defense mechanism against zooplankton and other grazers. Indeed, some zooplankton can become slowly incapacitated while feeding, as though they are being gradually paralyzed or otherwise impaired. (In one study, a tintinnid ciliate could swim only backward, away from its intended prey, after exposure to toxic dinoflagellates.) Sometimes grazing animals spit out the toxic algae as though they had an unpleasant taste. These responses would all reduce grazing and thus facilitate bloom formation.

All the same, nontoxic phytoplankton also form blooms, and so it is unlikely that toxins serve solely as self-defense. Scientists are looking within the algae for biochemical pathways that require the toxins, but the search thus far has been fruitless. The toxins are not proteins, and all are synthesized in a series of chemical steps requiring multiple genes. Investigators have proposed biosynthetic pathways, but they have not isolated chemical intermediates or enzymes used only in toxin production. It has thus been difficult to apply the powerful tools of molecular biology to these organisms, other than to study their genes or to develop detection tools.

We do have some tantalizing clues about toxin metabolism. For example, certain dinoflagellate strains produce different amounts of toxin and different sets of toxin derivatives when we vary their growth conditions. Metabolism of the toxins is a dynamic process, but we still do not know whether they have a specific biochemical role. As with the spiny diatoms that kill fish, the illnesses and mortalities caused by algal "toxins" may be the result of the accidental chemical affinity of those metabolites for receptor sites on ion channels in higher animals.

The potential role of bacteria or bacterial genes in phytoplankton toxin production is an area of active research. We wonder how a genetically diverse array of organisms, including phytoplankton, seaweeds, bacteria and cyanobacteria, could all have evolved the genes needed to produce saxitoxin [see "The Toxins of Cyanobacteria," by Wayne W. Carmichael; *Scientific American*, January 1994]. Several years ago Masaaki Kodama of Kitasato University in Japan isolated intracellular bacteria from antibiotic-treated *A. tamarense* cultures and showed that the bacteria produced saxitoxin. This finding supported an old and long-ignored hypothesis that toxins might originate from bacteria living inside or on the dinoflagellate cell.

Despite considerable study, the jury is still out. Many scientists now accept that some bacteria produce saxitoxins, but they point out that dense bacterial cultures produce extremely small quantities. It is also not clear that those bacteria can be found inside dinoflagellates. That intracellular bacteria produce all of the toxin found in a dinoflagellate cell therefore seems unlikely, but perhaps some synergism occurs between a small number of symbionts and the host dinoflagellate that is lost when the bacteria are isolated in culture. Alternatively, a bacterial gene or plasmid might be involved.

Given the diverse array of algae that produce toxins or cause problems in a variety of oceanographic systems, attempts to generalize the dynamics of harmful algal blooms are doomed to fail. Many harmful species, however, share some mechanisms. Red tides often occur when heating or freshwater runoff creates a stratified surface layer above colder, nutrient-rich waters. Fast-growing algae quickly strip away nutrients in the upper layer, leaving nitrogen and phosphorus only below the interface of the layers, called the pycnocline. Nonmotile phytoplankton cannot easily get to this layer, whereas motile algae, including dinoflagellates, can thrive. Many swim at speeds in excess of 10 meters per day, and some undergo daily vertical migration: they reside in surface waters by day to harvest sunlight like sunbathers, then swim down to the pycnocline to take up nutrients at night. As a result, blooms can suddenly appear in surface waters that are devoid of nutrients and would seem incapable of supporting such prolific growth.

A similar sleight-of-hand can occur horizontally, though over much larger distances. The NSP outbreak in North Carolina illustrates how ocean currents can transport major toxic species from one area to another. Patricia A. Tester, a biologist at the National Oceanic and Atmospheric Administration's National Marine Fisheries Service laboratory in Beaufort, examined plankton from local waters under a microscope soon after the initial reports of human illnesses. She saw cells resembling the dinoflagellate G. breve, the cause of recurrent NSP along Florida's western coast. Experts quickly confirmed her tentative identification, and for the first time in state history, authorities closed shellfish beds because of algal toxins, resulting in a loss of $20 million.

Tester and her co-workers have since used satellite images of sea-surface temperatures to argue that the G. breve population in North Carolina originated off the southwestern coast of Florida, nearly 1,000 kilometers away. That bloom traveled from the Gulf of Mexico up the southeastern coast of the U.S., transported by several current systems culminating in the Gulf Stream. After 30 days of transport, a filament of water separated from the Gulf Stream and moved onto North Carolina's narrow continental shelf, carrying *G. breve* cells with it. The warm water mass remained in nearshore waters, identifiable in satellite images for three weeks. Fortunately, *G. breve* does not have a known cyst stage, so it could not establish a seedbed and colonize this new region.

This incident, taken together with many others like it throughout the world, speaks of an unsettling trend. Problems from harmful red tides have grown worse over the past two decades. The causes, however, are multiple, and only some relate to pollution or other human activities. For example, the global expansion in aquaculture means that more areas are monitored closely, and more fisheries' products that can be killed or take up toxins are in the water. Likewise, our discovery of toxins in algal species formerly considered nontoxic reflects the maturation of this field of science, now profiting from more investigators, better analytical techniques and chemical instrumentation, and more efficient communication among workers.

Long-term studies at the local or regional level do show that red tides (in the most general sense of the term) are increasing as coastal pollution worsens. Between 1976 and 1986, as the population around Tolo Harbor in Hong Kong grew sixfold, red tides increased eightfold. Pollution presumably provided more nutrients to the algae. A similar pattern emerged in the Inland Sea of Japan, where visible red tides proliferated steadily from 44 per year in 1965 to more than 300 a decade later. Japanese authorities instituted rigorous effluent controls in the mid-1970s, and a 50 percent reduction in the number of red tides ensued.

These examples have been criticized, since both could be biased by changes in the numbers of observers through time, and both are tabulations of water discolorations from blooms, not just toxic or harmful episodes. Still, the data demonstrate what should be an obvious relationship: coastal waters receiving industrial, agricultural and domestic waste, frequently rich in plant nutrients, should experience a general increase in algal growth. These nutrients can enhance toxic or harmful episodes in several ways. Most simply, all phytoplankton species, toxic and nontoxic, benefit, but we notice the enrichment of toxic ones more. Fertilize your lawn, and you get more grass—and more dandelions.

Some scientists propose instead that pollution selectively stimulates harmful species. Theodore J. Smayda of the University of Rhode Island brings the nutrient ratio hypothesis, an old concept in the scientific literature, to bear on toxic bloom phenomena. He argues that human activities

Outbreaks of paralytic shellfish poisoning affected more than twice as many areas in 1990 as they did in 1970. Some experts believe coastal pollution and shipping practices have contributed to the expansion.

have altered the relative availability of specific nutrients in coastal waters in ways that favor toxic forms. For example, diatoms, most of which are harmless, require silicon in their cell walls, whereas other phytoplankton do not. Because silicon is not abundant in sewage, but nitrogen and phosphorus are, the ratio of nitrogen to silicon or of phosphorus to silicon in coastal waters has increased over the past several decades. Diatom growth ceases when silicon supplies are depleted, but other phytoplankton classes, which often include more toxic species, can proliferate using "excess" nitrogen and phosphorus. This idea is controversial but not unfounded. A 23-year time series from the German coast documents a fourfold rise in the nitrogen-silicon and phosphorus-silicon ratios, accompanied by a striking change in the composition of the phytoplankton community: diatoms decreased, whereas flagellates increased more than 10-fold.

Another concern is the long-distance transport of algal species in cargo vessels. We have long recognized that ships carry marine organisms in their ballast water, but evidence is emerging that toxic algae have also been hitchhiking across the oceans. Gustaaf M. Hallegraeff of the University of Tasmania has frequently donned a miner's helmet and ventured into the bowels of massive cargo ships to sample sediments accumulated in ballast tanks. He found more than 300 million toxic dinoflagellate cysts in one vessel alone. Hallegraeff argues that one PSP-producing dinoflagellate species first appeared in Tasmanian waters during the past two decades, concurrent with the development of a local wood-chip industry. Empty vessels that begin a journey in a foreign harbor pump water and sediment into their tanks for ballast; when wood chips are loaded in Tasmania, the tanks are discharged. Cysts easily survive the transit cruise and colonize the new site. Australia has now issued strict guidelines for discharging ballast water in the country's ports. Unfortunately, most other nations do not have such restrictions.

The past decade may be remembered as the time that humankind's effect on the global environment caught the public eye in a powerful and ominous fashion. For some, signs of our neglect come with forecasts of global warming, deforestation or decreases in biodiversity. For me and my colleagues, this interval brought a bewildering expansion in the complexity and scale of the red tide phenomenon. The signs are clear that pollution has enhanced the abundance of algae, including harmful and toxic forms. This effect is obvious in Hong Kong and the Inland Sea of Japan and is perhaps real but less evident in regions where coastal pollution is more gradual and unobtrusive. But we cannot blame all new outbreaks and new problems on pollution. There are many other factors that contribute to the proliferation of toxic species; some involve human activities, and some do not. Nevertheless, we may well be witnessing a sign that should not be ignored. As a growing world population demands more and more of fisheries' resources, we must respect our coastal waters and minimize those activities that stimulate the spectacular and destructive outbreaks called red tides.

FURTHER READING

PRIMARY PRODUCTION AND THE GLOBAL EPIDEMIC OF PHYTOPLANKTON BLOOMS IN THE SEA: A LINKAGE? Theodore J. Smayda in *Novel Phytoplankton Blooms: Causes and Impacts of Recurrent Brown Tide and Other Unusual Blooms.* Edited by E. M. Cosper, V. M. Bricelj and E. J. Carpenter. Springer-Verlag, 1989.

MARINE BIOTOXINS AT THE TOP OF THE FOOD CHAIN. Donald M. Anderson and Alan W. White in *Oceanus,* Vol. 35, No. 3, pages 55–61; Fall 1992.

DOMOIC ACID AND AMNESIC SHELLFISH POISONING: A REVIEW. Ewen C. D. Todd in *Journal of Food Protection,* Vol. 56, No. 1, pages 69–83; January 1993.

A REVIEW OF HARMFUL ALGAL BLOOMS AND THEIR APPARENT GLOBAL INCREASE. Gustaaf M. Hallegraeff in *Phycologia,* Vol. 32, No. 2, pages 79–99; March 1993.

MARINE TOXINS. Takeshi Yasumoto and Michio Murata in *Chemical Reviews,* Vol. 93, No. 5, pages 1897–1909; July/August 1993.

Natural Oil Spills

IAN R. MACDONALD

ORIGINALLY PUBLISHED IN NOVEMBER 1998

Beneath the Gulf of Mexico, to the south of Texas and Louisiana, tiny bubbles of oil and natural gas trickle upward through faulted marine sediments. Close to the seafloor, these hydrocarbons ooze past a final layer teeming with exotic deep-sea life before they seep into the ocean above. Buoyant, they rise through the water in tight, curving plumes, eventually reaching the surface. There the gas merges with the atmosphere, and the oil drifts downwind, evaporating, mixing with water and finally dispersing.

The best time to witness such a natural "oil spill" is in summer, when the Gulf stays flat calm for days at a time. In the middle of the afternoon, with the full heat of the tropical sun blazing off the sea, one can stand on the deck of a ship and watch broad ribbons of oil stretch toward the horizon. Cruising upwind along one of these slicks, one will notice that the sea takes on an unusual smoothness. The clarity of the water seems to increase, and the glare of the sun off the surface intensifies. Flying fish break from the bow waves and plunge into the water again almost without making a splash. Presently, the scent of fresh petroleum becomes evident—an odor that is quite distinct from the diesel fumes wafting from the ship— and one sees waxy patches floating on the water or clinging to the hull.

Abruptly, droplets of oil begin bursting into little circles of rainbow sheen, which expand rapidly from the size of a saucer to a dinner plate to a pizza pan and then disappear, merging with the general glassy layer on the water and drifting away. Continue on an upwind course, and the sea regains its normal appearance. The water darkens, breezes can once again raise a tracery of tiny wavelets, and flying fish make their usual splashes. The ship has sailed beyond the oil slick. But off in the distance lies another and another. I have heard Coast Guard pilots say that before they knew better, they wasted hours flying up such slicks in search of a vessel spewing oil.

Indeed, the ongoing release of hydrocarbons from the seabed creates slicks that closely resemble the notorious results of surreptitious bilge pumping. Yet discharges of oil from the deep are a natural consequence of the geologic circumstances that make the Gulf of Mexico one of the great hydrocarbon basins of the world.

TIME AND TIDE

The oil that leaks upward from the bottom of the Gulf—like oil found everywhere—forms because geothermal energy constantly bakes the organic matter buried within sedimentary rock. Over time, the hydrocarbons created in this way rise from deeper layers until they become trapped in porous sandstones, fractured shales or the limestone remnants of ancient reefs.

Apart from having abundant source rocks and plentiful geologic "traps" for rising oil, the Gulf is also special because it contains an ancient salt bed, which was laid down during repeated episodes of evaporation in the Jurassic period, about 170 million years ago. This layer, known as the Louann Salt, underlies most oil fields in the region. The crystalline salt is malleable but relatively incompressible. Over geologic time, the weight of accumulated sediment—much of it transported offshore from land—has tended to force the salt upward and outward, forming sheets, spires or ridges. Some of these structures retain contact with the parent bed; others move as separate bodies through the surrounding sediment.

This so-called salt tectonism affects the migration of hydrocarbons in a number of ways. For example, salt is impermeable and can readily trap hydrocarbons below it. Also, the movement of salt can open large faults that extend from deeply buried reservoirs all the way to the surface, providing conduits through which petroleum can travel upward.

The presence of such structures makes the Gulf of Mexico a unique place, one that looms very large in the history of the oil industry. Offshore oil and gas production was invented in the Gulf when the first platform was installed south of Louisiana in 1947. In subsequent years, operations moved farther and farther offshore as engineering advances made it possible to find and extract oil from below ever greater depths of water.

The social and economic consequences of this expansion have been pervasive. It is now almost impossible to imagine the Gulf coast without its population of oil workers, drilling rigs, production platforms, pipelines, tankers and refineries. To a degree not equaled elsewhere in the world, the Gulf of Mexico is a place where people live on and work under the sea—all to help satisfy society's insatiable hunger for petroleum.

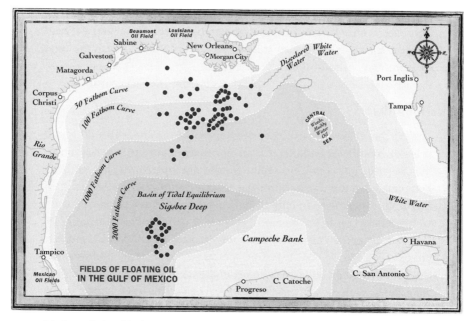

Soley's compilation of oil slicks observed between 1902 and 1909 (*dots above*) shows a preponderance of sightings southwest of the Mississippi Delta. The position of seafloor hydrocarbon seeps recently found in the Gulf of Mexico using modern methods suggests that many of the seeps active of the turn of the century are still discharging oil today.

Natural oil seeps offer a fascinating perspective on this enterprise. Although their existence comes as a surprise to many people, these seeps are well documented in the historical record. Pre-Columbian artifacts from the region show that tribal peoples commonly used beach tar as a caulking material, and Spanish records of floating oil date from the 16th century. In 1910 Lieutenant John C. Soley of the U.S. Navy published the first systematic study of offshore slicks in, curiously enough, *Scientific American* [see "The Oil Fields of the Gulf of Mexico," by John C. Soley; *Scientific American Supplement,* No. 1788, April 9, 1910].

Soley reviewed sightings of oil at sea that had been noted in the handwritten logs of ships. A report from the steamship *Comedian,* for example, described oil coming up in three jets at one location. These accounts were written many decades before the first offshore platforms were built, at a time that even predates the widespread use of petroleum as a shipping fuel. So the source of all this oil was something of an enigma.

Soley's theories about the origin of the oil seem naive, in their details at least, in light of modern petroleum geology. For example, he thought that much of the petroleum found far from shore was inorganic. Still, the details recorded by these unnamed mariners—the floating mats of oil, fouled seabirds, lumps of tar and acrid stench of petroleum—evoke the all too familiar images of modern oil spills. In fact, the descriptions of turn-of-the-century oil slicks are so dramatic that one wonders whether the amount of seepage going on then was greater than it is today.

SEEKING SEEPS

Soley had to rely on anecdotal descriptions and imperfect navigation to map oil seeps. He probably could not have anticipated that the physical and chemical properties of floating oil make it visible from such great distances that scientists can now monitor it over vast areas.

When thicker than four microns—the magnitude that might result from a shipping accident—fresh oil forms an obvious covering, reddish brown to tan in color. With exposure, the volatile fraction of the oil rapidly evaporates, leaving a waxy residue that makes a foamy emulsion with seawater and tends to coagulate into gooey tar balls and floating mats. In thicknesses between one and four microns, an oily layer refracts incident light to form the rainbow sheen familiar from curbside puddles. Natural oil slicks—which range from less than 0.01 to one micron—may be only a few tens of molecules thick. Still, the chemical bonds between hydrocarbons are sufficient to create a surprisingly durable film. This surface-active, or "surfactant," layer suppresses the fine-scale ripples that the wind would otherwise raise. The lack of ripples in turn allows the water to reflect light almost as effectively as a mirror, which gives a patch of floating oil its characteristic slick appearance.

The contrast between the reflection from a thin veneer of oil and the normal scattering of light from seawater makes these slicks quite distinctive under certain conditions. With the sun at a favorable angle, the slicks visible from a ship are also readily seen from airplanes and even from orbiting spacecraft.

Indeed, astronauts riding in the space shuttle can see slicks readily when they view the glint of the sun on the sea, a tactic they use to study ocean currents. In the center of a sun-glint pattern, the glare from oily patches is considerably brighter than the more diffuse reflection from unaffected waters. The situation reverses at the edges of the scene, where the geometry of illumination tends to direct light rays away from the viewer so that slicks look darker than their surroundings.

Slicks also appear dark under radar "illumination" because the source of the radio beam is in the same position as the detector, generally oblique to the water. The advent of satellite-mounted radars such as the European

SUN

PATCH APPEARS
BRIGHT

PATCH APPEARS
DARK

Sun-glint pattern seen by astronauts in the space shuttle Atlantis in orbit over the Gulf of Mexico displays many separate oil slicks. Near the center of the glow, where the solar reflection is most intense,
the slicks look especially bright. But at the periphery of the sun-glint pattern, the slicks appear dark. This
difference arises because the mirrorlike surface of the slick reflects rays toward the viewer's eye in the
first instance and away from it in the second, whereas the rougher surface of the unoiled water, which
tends to scatter the incident light in both cases, appears more uniform.

Radar Satellite, the Canadian RADARSAT and the space shuttle radar has
meant that almost any location in the ocean can be monitored for traces
of oil. Scientists sometimes use radar reflections from natural oil slicks to
study how seawater circulates. The results they obtain can show details
that would be difficult to discern with conventional oceanographic instruments. And geologists seeking telltale signs of hidden oil reservoirs
can also take advantage of the views attainable from space.

SORTING WHEAT FROM CHAFF

Using remote sensing to study natural oil seeps requires some way to distinguish the thin layers they generate from the vast majority of surfactant
slicks, which have nothing to do with petroleum. Fish spawn, plant waxes
and plankton swarms, among other phenomena, produce surfactants that,
when concentrated by converging currents, can generate detectable slicks.
But the markings of natural seepage are easily recognizable.

Typically, oil and gas leak from a group of vents spaced along a few hundred meters of a fault segment on the seafloor. Thus, the source is essentially fixed. Mathematical models indicate that a stream of oil drops rising
from a single orifice through a kilometer of water will surface within a
relatively small area. This theoretical result describes just what happens
in the Gulf of Mexico, where the rainbow traces of oil surfacing from a
vent at the head of a slick occupy an area no more than 100 meters or so
across. Over hours or days, currents at different depths will move this oily
footprint around somewhat. But generally, the oil reaches the surface less
than a kilometer or two from the underlying vent. So in practice, the repeated detection of a slick emanating from a stable location within an
area of that size points to a source located on the seafloor.

This approach requires that one can tell the head of a slick from its tail. And here intuition is misleading. Familiar analogues, such as a plume of smoke rising from a cigarette, might suggest that the skinny end marks the origin. But the situation is fundamentally different. Whereas smoke particles are unlikely to reassemble themselves into a concentrated mass, spilled oil coagulates, forming the many individual bands that can be seen streaming away from a cluster of sources on the seabed.

At the start of each slick, oil drops expand after surfacing. Simultaneously, the oil drifts under the influence of the wind and the current. In principle, an oil drop floating on water should spread outward until it forms a layer that is only one molecule thick. In actuality, the edges of a band of oil taper to the point where the slick cannot sustain itself. Downstream from the source, the layer becomes quite thin (on the order of 0.1 micron), but it is considerably thicker than the dimensions of a single molecule. Individual bands remain distinct for some distance, then merge and finally disappear. The overall outline of the slick is thus broader at its source than at its termination.

The length of a slick depends on the sea state. In heavy seas, an oily layer breaks up fairly rapidly and cannot drift far. Yet on a calm day, a layer of oil can be visible for 25 kilometers. Bends in the path traced by a slick reflect wind patterns: broad curves indicate gradual changes in wind direction, whereas chevrons document abrupt shifts. The time elapsed between a sudden reversal in wind direction and the moment a satellite acquires an image showing the resulting deviation indicates the life span of the slick. Using such comparisons, my co-workers at Texas A&M University and I typically find that the oil at the end of a visible slick has been floating on the surface for approximately 12 to 24 hours.

LIVING OFF OIL

Natural seeps may be a boon for oceanographers charting subtle eddies or for oil companies looking for new deposits to tap, but are they a bane for marine life? When the existence of the Gulf of Mexico oil seeps began to be widely recognized during the 1980s, my colleague Mahlon C. Kennicutt and his fellow researchers at Texas A&M speculated that the fauna living around seeps would provide a natural analogue for marine life exposed to oil pollution. To collect some of these presumably diseased specimens, they dragged a fishing net over active seeps. One of their first hauls, when it was brought on board their vessel, contained more than 800 kilograms of an unusual species of clam, *Calyptogena ponderosa*. Strangely, this large and obviously thriving creature was recovered from depths where deep-sea life normally proves rather scarce.

Adding to the mystery were dozens of brown, fibrous stalks also found in the net. These objects were so unfamiliar that the researchers almost threw them overboard, under the impression that the slender masses were merely some sort of reedy plant that had washed down the Mississippi River and settled in the deep Gulf. But one of the crew thought that this material might be good for weaving baskets. In sorting the stalks, he broke several of them open and spilled their red blood onto the deck, alerting the team that they had discovered something unexpected.

The specimens were eventually sent off to experts around the world. In the following months and years, a remarkable story began to emerge. Hydrocarbons leaking from the seafloor provide a source of chemical energy that nourishes creatures similar to the fauna first found at the hydrothermal vents of the Pacific Ocean in 1977. Vestimentiferan tube worms (the "stalks"), giant clams and a certain kind of deep-sea mussel make a living in both habitats through symbiosis with bacteria. Living within the cells of the animals, these bacteria synthesize new organic material in the absence of sunlight, using energy gained from oxidation of reduced compounds such as hydrogen sulfide.

The netting of vent creatures from the bottom of the Gulf of Mexico prompted investigators to take a closer look at this strange environment. In 1986 I led the expedition that made the first dives to an oil seep in a research submarine. My colleagues and I fully expected that we would have to search hard to find a few tube worms eking out a marginal existence. Instead we dropped right into the middle of a lush seafloor habitat, where we encountered large beds of mussels clustered around bubbling gas vents and extensive mats of brightly colored bacteria. Feeding on these exotic species was a diverse assemblage of fishes, crustaceans and other invertebrates that are commonly found in smaller numbers at shallower depths.

We now know that such thriving communities exist in many places in the Gulf. Interestingly, some of the biological activity at natural oil seeps tends to plug up the pores and fissures there. The metabolic by-products of microbes, in particular, cause the precipitation of calcium carbonate, which sometimes produces massive pavements that can trap oil below them. The creation of so-called gas hydrate on the ocean bottom can also block active gas vents.

Gas hydrate is an icelike substance that forms under conditions of high pressure and low but above-freezing temperature, when molecules of methane or other gases become trapped in a lattice of water molecules. Gas hydrates received early attention when these icy solids obstructed gas pipelines, forcing offshore operators to spend millions of dollars heating and insulating their undersea plumbing. More recently, researchers have focused on the gas hydrates that crystallize under the seafloor.

NATURE'S POLLUTION?

How much oil seeps naturally into the Gulf of Mexico? Assuming, conservatively, that an individual slick is 100 meters wide and maintains an average thickness of 0.1 micron over 10 kilometers, it must contain about 100 liters of oil. The life span of such a slick is typically 24 hours or less, indicating that while active, its source must have released at least 100 liters of oil a day. Estimating, conservatively again, that there are at least 100 such seeps in the Gulf at any time, then almost 40 million liters flow into the Gulf every decade.

It gives one pause to recall that the grounding of the *Exxon Valdez,* the benchmark of oil spills, dumped roughly the same amount of oil into Alaska's Prince William Sound. But it is staggering to consider that the releases in the Gulf of Mexico have been going on for a million years or more. Clearly, the ecosystem there has been able to cope with chronic oil "pollution" since long before the term was invented.

Comparing the natural release of oil through faults and fissures to its accidental release in the course of drilling or transport can be quite instructive—both for where the analogies hold and for where they break down. In the Gulf of Mexico, and probably in other parts of the world as well, natural leakage has extracted at least as much oil and gas from buried reservoirs as the petroleum industry has. But even though the total quantities may be roughly equivalent, the rates are not. Compared to nature, humankind is in a terrible hurry to get oil out of the ground.

That difference explains why a natural seep is not equivalent to a tanker accident, although the dosage might be identical. Just as a person who showers every day for a year may suffer the same exposure to water as one who drowns in a swimming pool, it is clear where the harm lies. The fact of natural oil seepage in no way forgives oil pollution. Whereas the open sea may tolerate hundreds of tons of oil a month—if it is spread thinly over tens of thousands of square kilometers—the same amount dumped on a seabird nesting area can obliterate the local population. Likewise, the chronic release of oil into an estuary may overwhelm that ecosystem.

Scientists and environmentalists alike must recognize that some oil spills can be quite damaging but that others are a benign part of the natural marine environment. The trick is to distinguish one from the other and to react appropriately. We do not want to be like the uninformed pilots, wasting time and energy just to chase sheen.

FURTHER READING

REASSESSMENT OF THE RATES AT WHICH OIL FROM NATURAL SOURCES ENTERS THE MARINE ENVIRONMENT. K. A. Kvenvolden in *Marine Environmental Research,* Vol. 10, pages 223–243; 1983.

NATURAL OIL SLICKS IN THE GULF OF MEXICO VISIBLE FROM SPACE. I. R. MacDonald et al. in *Journal of Geophysical Research,* Vol. 98, No. C9, pages 16351–16364; September 15, 1993.

REMOTE SENSING INVENTORY OF ACTIVE OIL SEEPS AND CHEMOSYNTHETIC COMMUNITIES IN THE NORTHERN GULF OF MEXICO. I. R. MacDonald et al. in *Hydrocarbon Migration and Its Near-Surface Expression.* Edited by D. Schumacher and M. A. Abrams. American Association of Petroleum Geologists Memoir 66, 1996.

Flammable Ice

ERWIN SUESS, GERHARD BOHRMANN, JENS GREINERT
AND ERWIN LAUSCH

ORIGINALLY PUBLISHED IN NOVEMBER 1999

I t was a thrilling moment when the enormous seafloor sampler opened
its metallic jaws and dumped its catch onto the deck of our ship, *Sonne*.
A white substance resembling effervescent snow gleamed amid the dark
mud hauled up from the bottom of the North Pacific Ocean. Watching it
melt before our eyes, we sensed that we had struck our own kind of gold.

As members of the Research Center for Marine Geosciences (GEOMAR)
at Christian Albrechts University in Kiel, Germany, we and our colleagues
were searching for methane hydrate—a white, icelike compound made
up of molecules of methane gas trapped inside cages of frozen water. To
that end, we had undertaken several expeditions to inspect, with the help
of a video camera tethered to the ship, a submarine ridge about 100 kilometers (62 miles) off the coast of Oregon. Earlier seismic investigations
and drilling had suggested that this area might hold a substantial stash
of our treasure. On July 12, 1996, we noticed peculiar white spots in the
mud 785 meters (2,575 feet) below our ship.

To make sure this telltale sign of hydrate was the real thing, we directed our sampler, a contraption like a backhoe with two scoops, to take
a giant bite out of the seafloor. Even while retrieving the payload, we saw
our expectations confirmed. As the sampler ascended, the video camera
mounted inside its jaws revealed that bubbles—attesting to the rapid escape of methane gas—were beginning to emerge from the muddy heap.
Stable only at near-freezing temperatures and under the high-pressure
conditions generated by the weight of at least 500 meters of overlying water, methane hydrates decompose rapidly above that depth. As the sample
approached the ocean's surface, the flow of bubbles gradually increased
and burst through the water's sparkling surface long before the jaws of
the sampler did.

We wondered how much intact hydrate would reach the deck. Moving
quickly, we managed to safeguard roughly 100 pounds (45 kilograms) of
the hissing chunks in containers cooled with liquid nitrogen. In the end,

we even had a few pieces left over, which inspired an impromptu fire-works display. Just holding a burning match to one of the white lumps ig-nited the hydrate's flammable methane, which also is one of the primary hydrocarbon components of natural gas. The lump burned with a reddish flame and left only a puddle of water as evidence of its former glory.

Before 1970 no one even knew that methane hydrates existed under the sea, and our haul was by far the largest quantity ever recovered from the ocean depths. Yet hydrates are by no means a rarity. On the contrary, in recent years they have been found to occur worldwide—from Japan to New Jersey and from Oregon to Costa Rica—in enormous quantities. Es-timates vary widely, but most experts agree that marine gas hydrates col-lectively harbor twice as much carbon as do all known natural gas, crude oil and coal deposits on the earth [*see illustration on page 270*].

The energy stored in methane hydrates could potentially fuel our energy-hungry world in the future (if practical mining techniques are de-vised). But the hydrates also have a worrisome aspect: methane escaping from disturbed undersea hydrates may be an ecological threat. If even a small portion of these deposits decompose through natural processes, as-tonishing quantities of methane will be set free to exacerbate the green-house effect and global warming.

Although methane remains in the atmosphere relatively briefly—10 years on average—it does not vanish without a trace. In the presence of free oxygen, a methane molecule's single atom of carbon disengages from its four hydrogen atoms to become carbon dioxide, the most infa-mous of all greenhouse gases because it is one of those spewed into the atmosphere during the combustion of fossil fuels.

But are decomposing methane hydrates contributing to global warm-ing now? And are they likely to do so in the future? Our 1996 journey—along with dozens of voyages and experiments since—have revealed some-thing about the structure and origins of a variety of these massive yet remarkably unstable deposits and have provided some answers to the cli-mate questions, but our understanding is far from complete. We and our colleagues at GEOMAR continue our quest to understand just what role methane hydrates play in ocean-floor stability and both past and future climatic change.

TURNING HEADS

It was the immense cache of energy trapped in marine methane hydrates that first turned the heads of politicians. But the challenges of tapping that resource are now making some officials look the other way. Hydrates tend to form along the lower margins of continental slopes, where the seabed drops from the relatively shallow shelf, usually about 150 meters below the surface, toward the ocean's abyss several kilometers deep. The hydrate deposits may reach beneath the ocean floor another few hundred meters—deeper than most drilling rigs can safely operate. Moreover, the roughly sloping seafloor makes it difficult to run a pipeline from such deposits to shore.

Countries that wish to rely less on foreign fossil fuels have started to overcome these technical difficulties, however. Japan was scheduled to launch an experimental hydrate drilling project off the coast of Hokkaido in October. U.S. engineers also are playing with ideas for tapping hydrate energy sources. But as long as relatively cheap gas and oil remain available, most industrial countries are unlikely to invest heavily in the technologies needed to harvest hydrates efficiently.

Methane hydrates captured the attention of petroleum geologists a bit earlier than that of politicians. When engineers first realized in the 1930s that gas-laden ice crystals were plugging their gas and oil pipelines, laboratory researchers spent time studying hydrate structure and composition. For example, they learned that one type of hydrate structure consists of icy cages that can absorb small gas molecules such as methane, carbon dioxide and hydrogen sulfide. A different type forms larger cavities that can enclose several small molecules or larger hydrocarbon molecules, such as pentane. What is more, the individual cages can differ in the kinds of gas molecules they capture.

In the 1960s scientists discovered that hydrates could also form in natural environments. They found the first natural deposits in the permafrost regions of Siberia and North America, where the substances were known as marsh gas. In the 1970s geophysicists George Bryan and John Ewing of Lamont-Doherty Earth Observatory of Columbia University found the earliest indication that methane hydrates also lurk beneath the seafloor. The hint came from seismological studies at Blake Ridge, a 100-kilometer-long feature off the North Carolina coast.

Seismologists can distinguish layers beneath the seafloor because sound waves bounce off certain kinds of dirt and rock differently than off other kinds. Some 600 meters below the ocean floor Bryan and Ewing saw an unusual reflection that mimicked the contour of the ridge. Their conclusion: this bottom-simulating reflector was the boundary between a methane hydrate layer and a layer of free methane gas that had accumulated below. Other experts found similar features elsewhere, and soon this type of reflector was being mapped as a methane hydrate deposit in ocean basins around the globe.

We used a bottom-simulating reflector, along with underwater video cameras, to guide our 1996 search for methane hydrates along the North Pacific seafloor promontory that has since been named Hydrate Ridge. Our successful recovery of intact hydrate on that expedition made it possible to study this unusual material in detail for the first time. Being able to analyze the texture and chemistry of its microscopic structure allowed us to confirm the plausible but previously unproved notion that the methane derives from the microbial decomposition of organic matter in the sediment.

Most telling were chemical tests that showed the hydrates to be enriched in carbon 12. Inorganic methane that seeps out of volcanic ridges and vents has higher levels of carbon 13, an isotope of carbon with an additional neutron. But the bacteria that digest organic matter in oxygen-deficient conditions such as those in sediments at the bottom of the sea tend to sequester more carbon 12 in the methane they generate.

The seafloor off the coast of Oregon also proved an especially fruitful theater of operations for assessing the stability of methane hydrate deposits and their potential role in releasing carbon into the atmosphere. Combined with research from other sites, these analyses indicate that methane hydrate deposits can be disturbingly labile.

We now know that in places along Hydrate Ridge, the seabed is virtually paved with hundreds of square meters of hydrate. These deposits form part of the packet of sediments riding piggyback on the Juan de Fuca tectonic plate, which is sliding underneath North America at a rate of 4.5 centimeters (1.7 inches) a year. As the Juan de Fuca plate is subducted, the sediments and hydrates it carries are partially sheared off by the upper plate and pressed into folds or piled several layers high like a giant stack of pancakes. This distorted material forms a wedge of mud that accumulates against the North American plate in the shape of ridges running nearly parallel to the coast.

METHANE PLUMES

In 1984 one of us (Suess) was the first to observe these ridges and their world of cold, eternal night from *Alvin*, the research submersible operated by the Woods Hole Oceanographic Institution. Outside *Alvin's* porthole, Suess saw a landscape of stone chimneys built by minerals precipitated from hazes of water and gas spewing out of the earth's crust. Only later did we realize that these chimneys are partially the product of a methane hydrate deposit being squeezed as one tectonic plate scrapes past the other.

Plumes of gas and fluids also escape along faults that cut through the sediment and gas hydrates alike. Although these plumes, also called cold vents, are unlike the hot springs that form along mid-oceanic ridges (where hot lava billows out of a crack in the seafloor), they are nonetheless warm enough to further destabilize the hydrates. These melt when the surrounding temperature creeps even a few degrees above freezing. As new hydrates form near the seafloor, the lower portions melt away, with the result that the overall layer migrates upward over time.

Melting at the bottom of a hydrate layer liberates not only freshwater but also methane and small amounts of hydrogen sulfide and ammonia. Oxidation of these chemicals into carbon dioxide, sulfate and nitrate provides nourishment to rich communities of chemical-eating bacteria. These microbes in turn serve as food for such creatures as clams and colonies of tube worms. Such oases of life stand out on the otherwise sparsely inhabited seabed.

Our investigations also revealed that the gases liberated at these densely populated vents give rise to an immensely active turnover of carbon. Oxidation of the liberated methane generates bicarbonate, which combines with calcium ions in the seawater to form calcium carbonate, better known as limestone. Such limestone is what Suess saw in 1984 in the form of chimneys and vent linings along the crest of Hydrate Ridge—and what we now realize is a hint of a deeper hydrate layer.

Along the western flank of Hydrate Ridge, massive limestone blocks cover the crack created by a large fault. But despite the limestone casing and the activities of the vent organisms, surprising quantities of methane escape into the surrounding ocean water. In fact, we measured concentrations that are roughly 1,300 times the methane content of water at equilibrium with the methane content of the air. We still do not know how much of the methane is oxidized in the water and how much actually enters the air, but it is easy to imagine that an earthquake or other dramatic tectonic event could release large amounts of this highly potent greenhouse gas into the atmosphere.

Researchers at GEOMAR have a much better idea of how much methane escapes in plumes rising up from hydrate fields in the Sea of Okhotsk off the east coast of Asia. About as big as the North Sea and the Baltic Sea combined, this body of water is delineated by the Kamchatka Peninsula and the Kuril island arc. In the summer of 1998 a joint German-Russian team

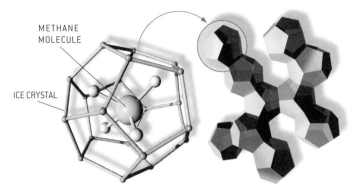

METHANE
MOLECULE

ICE CRYSTAL

Crystalline cages of frozen water sometimes snare molecules of methane gas that have been given off as microbes digest organic matter in the seafloor mud.

using fish-finder sonar documented methane plumes as tall as 500 meters billowing out of methane hydrate deposits on the seafloor. With our video camera tethered to the boat, we also saw giant chimneys reminiscent of the cold vents along Hydrate Ridge in the Pacific.

Even before we had visual evidence of the hydrate deposits, we knew that enormous quantities of methane accumulate under the blanket of ice that typically covers much of the Sea of Okhotsk for seven months a year. We measured a concentration of 6.5 milliliters of methane per liter of water just beneath the ice during a 1991 expedition. When the sea was free of ice the next summer, this figure was only 0.13—the difference had vented into the atmosphere. No similar methane flux has yet been observed anywhere else in the world, so this event may be unique. Still, this one-time measurement of methane escape from the Sea of Okhotsk clearly demonstrates that methane hydrates below the oceans can be a significant source of atmospheric methane. To help evaluate the possible current and future climatic impact of the methane, scientists are now sampling methane concentrations in the Okhotsk seawater every two months.

SHAKY GROUND

Plumes caused by seafloor faulting and the natural decomposition of hydrates can release methane slowly to the atmosphere, but it turns out that this process is sometimes much more explosive. In the summer of 1998 Russian researchers from the Shirshov Institute of Oceanology in Moscow found unstable hydrate fields off the west coast of Norway that they suspect are the cause of one of history's most impressive releases of trapped methane, an event known as the Storrega submarine landslide.

From previous explorations of the seafloor, scientists know that 8,000 years ago some 5,600 cubic kilometers (1,343 cubic miles) of sediments slid a distance of 800 kilometers from the upper edge of the continental slope into the basin of the Norwegian Sea, roughly at the latitude of Trondheim. The consequence of so much mud pushing water out of its path would have been devastating tsunamis—horrific swells that suddenly engulf the coastline.

The presence of methane hydrate fields in the same seafloor vicinity implies that unstable hydrates triggered the slide as they rapidly decomposed because of a change in the pressure or temperature after the last ice age. As the glaciers receded, the seafloor no longer had to support the enormous weight of the ice. As the land rebounded, the overlying sea and ice both warmed and became more shallow, suddenly moving the hydrates out of their zone of stability.

Could a geologic event like this strike again? Off the coast of southern Norway the risk of new slides appears to be relatively small, because the hydrate fields have for the most part decomposed. But the issue of the stability of the continental slope is assuming a heightened importance in view of current global warming and the strong possibility of further changes in the earth's climate in the near future. Beyond contributing to tsunamis, hydrate formations that become unstable and decompose will release methane into the oceans. In fact, melting a mere cubic meter of hydrate releases up to 164 cubic meters of methane, some of which would surely reach the atmosphere. In turn, a warming of the lower atmosphere would heat the oceans, launching a vicious circle of more dissolution of hydrates and more atmospheric warming.

Many researchers think that an explosive methane release from a single large site can create dramatic climate changes on short timescales. James P. Kennett, an oceanographer at the University of California at Santa Barbara, has hypothesized that catastrophic releases of methane could have triggered the notable increase in temperature that occurred over just a few decades during the earth's last ice age some 15,000 years ago. An international team led by former GEOMAR member Jürgen Mienert, now at Tromsø University in Norway, recently found possible evidence of this methane release on the floor of the Barents Sea, just off Norway's northeastern tip.

There, fields of giant depressions reminiscent of bomb craters pockmark the immediate vicinity of methane hydrate deposits. Mienert's team

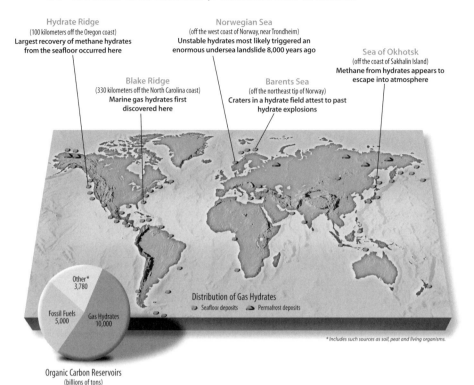

Hydrate Ridge
(100 kilometers off the Oregon coast)
Largest recovery of methane hydrates
from the seafloor occurred here

Norwegian Sea
(off the west coast of Norway, near Trondheim)
Unstable hydrates most likely triggered an
enormous undersea landslide 8,000 years ago

Sea of Okhotsk
(off the coast of Sakhalin Island)
Methane from hydrates appears to
escape into atmosphere

Blake Ridge
(330 kilometers off the North Carolina coast)
Marine gas hydrates first
discovered here

Barents Sea
(off the northeast tip of Norway)
Craters in a hydrate field attest to past
hydrate explosions

Other*
3,780

Fossil Fuels
5,000

Gas Hydrates
10,000

Distribution of Gas Hydrates
Seafloor deposits   Permafrost deposits

* Includes such sources as soil, peat and living organisms.

Organic Carbon Reservoirs
(billions of tons)

Hydrate deposits containing methane and other gases exist worldwide under the seafloor and in permafrost regions on land (*map*). Gas hydrates contain more organic carbon than does any other global reservoir (*pie chart*).

measured the biggest of these craters at 700 meters wide and 30 meters deep—a size clearly suggestive of catastrophic explosions of methane. Whether these eruptions occurred more or less simultaneously has not yet been determined, but faults and other structural evidence indicate they probably took place toward the end of the last ice age, as Kennett proposed. The explosions very likely followed a scenario like the one suggested for the cause of the Storrega landslide: warming seas rendered the hydrates unstable, and at a critical point they erupted like a volcano.

OLDER HINTS

Researchers have also uncovered evidence that methane liberated from gas hydrates affected the global climate in the more distant past—at the end of the Paleocene, about 55 million years ago. Fossil evidence suggests that land and sea temperatures rose sharply during this period. Many species of single-celled organisms dwelling in the seafloor sediment became extinct. A flux of some greenhouse gas into the atmosphere presumably warmed the planet, but what was the source? Carbon isotopes turned out to be the key to interpreting the cause of the rapid rise in temperature.

Scientists found a striking increase in the lighter carbon 12 isotope in the preserved shells of microscopic creatures that survived the heat spell. Methane hydrates are the likeliest source for the light carbon—and for the greenhouse gas flux—because these deposits are the only places where organic methane accumulates to levels that could influence the isotopic signature of the seawater when they melt. The carbon 12 enrichment characteristic of the hydrates disperses into the seawater with the liberated methane and persists in its oxidation product, carbon dioxide. Some of the carbon dioxide in turn becomes incorporated into the calcium carbonate shells of the sea creatures, while some of the methane makes its way to the atmosphere to help warm things up.

Gerald Dickens, now at James Cook University in Australia, used a computer simulation to test whether melting methane hydrates could have belched out enough greenhouse gases to subject the earth to a heat shock 55 million years ago. He and his former collaborators at the University of Michigan based their simulation on the assumption that hydrates corresponding to about 8 percent of today's global reserves decomposed at that time. Because liberated methane is converted immediately (on a geologic timescale) into carbon dioxide, they tracked this compound only.

In 10,000 simulated years, 160 cubic kilometers of carbon dioxide containing carbon 12 showed up in their model atmosphere every year. Adding this carbon dioxide caused the lower atmosphere to warm by two degrees Celsius (3.6 degrees Fahrenheit). At the same time, the isotope ratio of carbon in the water and atmosphere shifted to correspond to the values observed in the fossils. Moreover, this carbon isotope ratio gradually returned to normal within 200,000 years, just as it does in actual fossil records.

Dickens's model is compelling but rare. So far the significance of methane from natural gas hydrate sources as a greenhouse gas has received only limited consideration in global climate modeling. The contribution of methane hydrates to global carbon budgets has likewise not been adequately taken into account. Trying to come up with data to correct these shortfalls is one of the greatest motivations for our work on methane hydrates at GEOMAR. We continue to focus on Hydrate Ridge off the Oregon coast—eight marine expeditions targeted the site earlier this year.

We will also be looking at the seafloor off the coasts of Costa Rica, Nicaragua, Alaska's Aleutian Islands and New Zealand. Whatever happens in methane hydrate research, the future will be anything but dull.

FURTHER READING

AUTHIGENIC CARBONATES FROM THE CASCADIA SUBDUCTION ZONE AND THEIR RELATION TO GAS HYDRATE STABILITY. Gerhard Bohrmann et al. in *Geology*, Vol. 26, No. 7, pages 647–650; 1998.

FLUID VENTING IN THE EASTERN ALEUTIAN SUBDUCTION ZONE. Erwin Suess et al. in *Journal of Geophysical Research B: Solid Earth and Planets*, Vol. 103, No. B2, pages 2597–2614; February 10, 1998.

GAS HYDRATES: RELEVANCE TO WORLD MARGIN STABILITY AND CLIMATE CHANGE. Edited by J.-P. Henriet and J. Mienert. Geological Society Special Publications, Vol. 137; 1998.

POTENTIAL EFFECTS OF GAS HYDRATE ON HUMAN WELFARE. Keith A. Kvenvolden in *Proceedings of the National Academy of Sciences USA*, Vol. 96, No. 7, pages 3420–3426; March 30, 1999.

Can We Bury Global Warming?

ROBERT H. SOCOLOW

ORIGINALLY PUBLISHED IN JULY 2005

When William Shakespeare took a breath, 280 molecules out of every million entering his lungs were carbon dioxide. Each time you draw breath today, 380 molecules per million are carbon dioxide. That portion climbs about two molecules every year.

No one knows the exact consequences of this upsurge in the atmosphere's carbon dioxide (CO_2) concentration nor the effects that lie ahead as more and more of the gas enters the air in the coming decades—humankind is running an uncontrolled experiment on the world. Scientists know that carbon dioxide is warming the atmosphere, which in turn is causing sea level to rise, and that the CO_2 absorbed by the oceans is acidifying the water. But they are unsure of exactly how climate could alter across the globe, how fast sea level might rise, what a more acidic ocean could mean, which ecological systems on land and in the sea would be most vulnerable to climate change and how these developments might affect human health and well-being. Our current course is bringing climate change upon ourselves faster than we can learn how severe the changes will be.

If slowing the rate of carbon dioxide buildup were easy, the world would be getting on with the job. If it were impossible, humanity would be working to adapt to the consequences. But reality lies in between. The task can be done with tools already at hand, albeit not necessarily easily, inexpensively or without controversy.

Were society to make reducing carbon dioxide emissions a priority—as I think it should to reduce the risks of environmental havoc in the future—we would need to pursue several strategies at once. We would concentrate on using energy more efficiently and on substituting noncarbon renewable or nuclear energy sources for fossil fuel (coal, oil and natural gas—the primary sources of man-made atmospheric carbon dioxide). And we would employ a method that is receiving increasing attention: capturing carbon dioxide and storing, or sequestering, it underground rather than releasing it into the atmosphere. Nothing says that CO_2 must be emitted into the air. The atmosphere has been our prime waste repository, because discharging exhaust up through smokestacks, tailpipes and chimneys is the simplest and least (immediately) costly thing to do. The good news is that the technology for capture and storage already exists and that the obstacles hindering implementation seem to be surmountable.

CARBON DIOXIDE CAPTURE

The combustion of fossil fuels produces huge quantities of carbon dioxide. In principle, equipment could be installed to capture this gas wherever these hydrocarbons are burned, but some locations are better suited than others.

If you drive a car that gets 30 miles to the gallon and go 10,000 miles next year, you will need to buy 330 gallons—about a ton—of gasoline. Burning that much gasoline sends around three tons of carbon dioxide out the tailpipe. Although CO_2 could conceivably be caught before leaving the car and returned to the refueling station, no practical method seems likely to accomplish this task. On the other hand, it is easier to envision trapping the CO_2 output of a stationary coal-burning power plant.

It is little wonder, then, that today's capture-and-storage efforts focus on those power plants, the source of one quarter of the world's carbon dioxide emissions. A new, large (1,000-megawatt-generating) coal-fired power plant produces six million tons of the gas annually (equivalent to the emissions of two million cars). The world's total output (roughly equivalent to the production of 1,000 large plants) could double during the next few decades as the U.S., China, India and many other countries construct new power-generating stations and replace old ones [see illustration on page 276]. As new coal facilities come online in the coming quarter of a century, they could be engineered to filter out the carbon dioxide that would otherwise fly up the smokestacks.

Today a power company planning to invest in a new coal plant can choose from two types of power systems, and a third is under development but not yet available. All three can be modified for carbon capture. Traditional coal-fired steam power plants burn coal fully in one step in air: the heat that is released converts water into high-pressure steam, which turns a steam turbine that generates electricity. In an unmodified version of this system—the workhorse of the coal power industry for the past century—a mixture of exhaust (or flue) gases exits a tall stack at atmospheric pressure after having its sulfur removed. Only about 15 percent of the flue gas is carbon dioxide; most of the remainder is nitrogen and water vapor. To adapt this technology for CO_2 capture, engineers could replace the smokestack with an absorption tower, in which the flue gases would come in contact with droplets of chemicals called amines that selectively absorb CO_2. In a second reaction column, known as a stripper tower, the amine liquid would be heated to release concentrated CO_2 and to regenerate the chemical absorber.

The other available coal power system, known as a coal gasification combined-cycle unit, first burns coal partially in the presence of oxygen in a gasification chamber to produce a "synthetic" gas, or syngas—primarily pressurized hydrogen and carbon monoxide. After removing sulfur compounds and other impurities, the plant combusts the syngas in air in a gas turbine—a modified jet engine—to make electricity. The heat in the exhaust gases leaving the gas turbine turns water into steam, which is piped into a steam turbine to generate additional power, and then the gas turbine exhaust flows out the stack. To capture carbon from such a facility, technicians add steam to the syngas to convert (or "shift") most of the carbon monoxide into carbon dioxide and hydrogen. The combined cycle system next filters out the CO_2 before burning the remaining gas, now mostly hydrogen, to generate electricity in a gas turbine and a steam turbine.

The third coal power approach, called oxyfuel combustion, would perform all the burning in oxygen instead of air. One version would modify single-step combustion by burning coal in oxygen, yielding a fuel gas with no nitrogen, only CO_2 and water vapor, which are easy to separate. A second version would modify the coal gasification combined-cycle system by using oxygen, rather than air, at the gas turbine to burn the carbon monoxide and hydrogen mixture that has exited the gasifier. This arrangement skips the shift reaction and would again produce only CO_2 and water vapor. Structural materials do not yet exist, though, that can withstand the higher temperatures that are created by combustion in oxygen rather than in air. Engineers are exploring whether reducing the process temperature by recirculating the combustion exhaust will provide a way around these materials constraints.

TOUGH DECISIONS

Modification for carbon dioxide capture not only adds complexity and expense directly but also cuts the efficiency of extracting energy from the fuel. In other words, safely securing the carbon by-products means mining and burning more coal. These costs may be partially offset if the plant

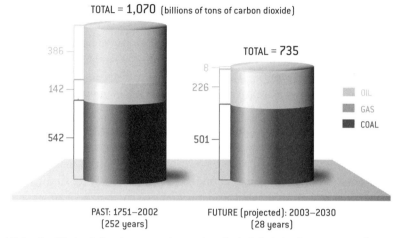

TOTAL = **1,070** (billions of tons of carbon dioxide)

TOTAL = **735**

386

142

542

8

226

501

OIL
GAS
COAL

PAST: 1751–2002   FUTURE (projected): 2003–2030
(252 years)   (28 years)

Lifetime fossil-fuel emissions from power plants projected to be built during the next quarter of a century will be comparable to all the emissions during the past 250 years. The left column shows the cumulative carbon dioxide emissions produced by burning coal, oil and natural gas for all uses [including
transportation and building heating] from 1751 to 2002, whereas that on the right depicts the lifetime
CO_2 emissions from fossil-fuel power generation plants projected by the International Energy Agency to
come online between 2003 and 2030. Coal-fired power plants are assumed to operate for 60 years and
gas-fired power stations for 40 years.

can filter out gaseous sulfur simultaneously and store it with the CO_2,
thus avoiding some of the considerable expense of sulfur treatment.

Utility executives want to maximize profits over the entire life of the
plant, probably 60 years or more, so they must estimate the expense of
complying not only with today's environmental rules but also with future
regulations. The managers know that the extra costs for CO_2 capture are
likely to be substantially lower for coal gasification combined-cycle plants
than for traditional plants. Removing carbon dioxide at high pressures,
as occurs in a syngas operation, costs less because smaller equipment can
be employed. But they also know that only a few demonstration gasification plants are running today, so that opting for gasification will require
spending extra on backup equipment to ensure reliability. Hence, if the
management bets on not having to pay for CO_2 emissions until late in the
life of its new plant, it will probably choose a traditional coal plant, although perhaps one with the potential to be modified later for carbon capture. If, however, it believes that government directives to capture CO_2 are
on their way within a decade or so, it may select a coal gasification plant.

To get a feel for the economic pressures the extra cost of carbon sequestration would place on the coal producer, the power plant operator and

the home owner who consumes the electricity, it helps to choose a reasonable cost estimate and then gauge the effects. Experts calculate that the total additional expense of capturing and storing a ton of carbon dioxide at a coal gasification combined-cycle plant will be about $25. (In fact, it may be twice that much for a traditional steam plant using today's technology. In both cases, it will cost less when new technology is available.)

The coal producer, the power plant operator and the home owner will perceive that $25 cost increase quite differently. A coal producer would see a charge of about $60 per ton of coal for capturing and storing the coal's carbon, roughly tripling the cost of coal delivered to an electric utility customer. The owner of a new coal power plant would face a 50 percent rise in the cost of power the coal plant puts on the grid, about two cents per kilowatt-hour (kWh) on top of a base cost of around four cents per kWh. The home owner buying only coal-based electricity, who now pays an average of about 10 cents per kWh, would experience one-fifth higher electricity costs (provided that the extra two cents per kWh cost for capture and storage is passed on without increases in the charges for transmission and distribution).

FIRST AND FUTURE STEPS

Rather than waiting for the construction of new coal-fired power plants to begin carbon dioxide capture and storage, business leaders are starting the process at existing facilities that produce hydrogen for industry or purify natural gas (methane) for heating and power generation. These operations currently generate concentrated streams of CO_2. Industrial hydrogen production processes, located at oil refineries and ammonia plants, remove carbon dioxide from a high-pressure mix of CO_2 and hydrogen, leaving behind carbon dioxide that is released skyward. Natural gas purification plants must remove CO_2 because the methane is heading for a liquefied natural gas tanker and must be kept free of cold, solid carbon dioxide (dry ice) that could clog the system or because the CO_2 concentration is too high (above 3 percent) to be allowed on the natural gas distribution grid.

Many carbon dioxide capture projects using these sources are now under consideration throughout the oil and gas industry. Hydrogen production and natural gas purification are the initial stepping-stones to full-scale carbon capture at power plants; worldwide about 5 percent as much carbon dioxide is produced in these two industries as in electric power generation.

In response to the growing demand for imported oil to fuel vehicles, some nations, such as China, are turning to coal to serve as a feedstock for synthetic fuels that substitute for gasoline and diesel fuel. From a climate change perspective, this is a step backward. Burning a coal-based synthetic fuel rather than gasoline to drive a set distance releases approximately double the carbon dioxide, when one takes into account both tailpipe and synfuels plant emissions. In synthetic fuels production from coal, only about half the carbon in the coal ends up in the fuel, and the other half is emitted at the plant. Engineers could modify the design of a coal synfuels plant to capture the plant's CO_2 emissions. At some point in the future, cars could run on electricity or carbon-free hydrogen extracted from coal at facilities where CO_2 is captured.

Electricity can also be made from biomass fuels, a term for commercial fuels derived from plant-based materials: agricultural crops and residues, timber and paper industry waste, and landfill gas. If the fossil fuels used in harvesting and processing are ignored, the exchanges between the atmosphere and the land balance because the quantity of carbon dioxide released by a traditional biomass power plant nearly equals that removed from the atmosphere by photosynthesis when the plants grew. But biomass power can do better: if carbon capture equipment were added to these facilities and the harvested biomass vegetation were replanted, the net result would be to scrub the air of CO_2. Unfortunately, the low efficiency of photosynthesis limits the opportunity for atmospheric scrubbing because of the need for large land areas to grow the trees or crops. Future technologies may change that, however. More efficient carbon dioxide removal by green plants and direct capture of CO_2 from the air (accomplished, for example, by flowing air over a chemical absorber) may become feasible at some point.

CARBON DIOXIDE STORAGE

Carbon capture is just half the job, of course. When an electric utility builds a 1,000-megawatt coal plant designed to trap CO_2, it needs to have somewhere to stash securely the six million tons of the gas the facility will generate every year for its entire life. Researchers believe that the best destinations in most cases will be underground formations of sedimentary rock loaded with pores now filled with brine (salty water). To be suitable, the sites typically would lie far below any source of drinking water, at least 800 meters under the surface. At 800 meters, the ambient pressure is 80 times that of the atmosphere, high enough that the pressurized injected CO_2 is in a "supercritical" phase—one that is nearly as dense as the brine it replaces in geologic formations. Sometimes crude oil or natural gas will also be found in the brine formations, having invaded the brine millions of years ago.

The quantities of carbon dioxide sent belowground can be expressed in "barrels," the standard 42-gallon unit of volume employed by the petroleum industry. Each year at a 1,000-megawatt coal plant modified for carbon capture, about 50 million barrels of supercritical carbon dioxide would be secured—about 100,000 barrels a day. After 60 years of operation, about three billion barrels (half a cubic kilometer) would be sequestered below the surface. An oil field with a capacity to produce three billion barrels is six times the size of the smallest of what the industry calls "giant" fields, of which some 500 exist. This means that each large modified coal plant would need to be associated with a "giant" CO_2 storage reservoir. About two thirds of the 1,000 billion barrels of oil the world has produced to date has come from these giant oil fields, so the industry already has a good deal of experience with the scale of the operations needed for carbon storage.

Many of the first sequestration sites will be those that are established because they can turn a profit. Among these are old oil fields into which carbon dioxide can be injected to boost the production of crude. This so-called enhanced oil recovery process takes advantage of the fact that pressurized CO_2 is chemically and physically suited to displacing hard-to-get oil left behind in the pores of the geologic strata after the first stages of production. In this process, compressors drive CO_2 into the oil remaining in the deposits, where chemical reactions result in modified crude oil that moves more easily through the porous rock toward production wells. In particular, CO_2 lowers crude oil's interfacial tension—a form of surface tension that determines the amount of friction between the oil and rock. Thus, carbon dioxide injects new life into old fields.

In response to British government encouragement of carbon dioxide capture and storage efforts, oil companies are proposing novel capture projects at natural gas power plants that are coupled with enhanced oil recovery ventures at fields underneath the North Sea. In the U.S., operators of these kinds of fields can make money today while paying about $10 to $20 per ton for carbon dioxide delivered to the well. If oil prices continue to rise, however, the value of injected CO_2 will probably go up because its use enables the production of a more valuable commodity. This market development could lead to a dramatic expansion of carbon dioxide capture projects.

Carbon sequestration in oil and gas fields will most likely proceed side by side with storage in ordinary brine formations, because the latter structures are far more common. Geologists expect to find enough natural storage capacity to accommodate much of the carbon dioxide that could be captured from fossil fuels burned in the 21st century.

STORAGE RISKS

Two classes of risk must be addressed for every candidate storage reservoir: gradual and sudden leakage. Gradual release of carbon dioxide merely returns some of the greenhouse gas to the air. Rapid escape of large amounts, in contrast, could have worse consequences than not storing it at all. For a storage operation to earn a license, regulators will have to be satisfied that gradual leakage can occur only at a very slow rate and that sudden leakage is extremely unlikely.

Although carbon dioxide is usually harmless, a large, rapid release of the gas is worrisome because high concentrations can kill. Planners are well aware of the terrible natural disaster that occurred in 1986 at Lake Nyos in Cameroon: carbon dioxide of volcanic origin slowly seeped into the bottom of the lake, which sits in a crater. One night an abrupt overturning of the lake bed let loose between 100,000 and 300,000 tons of CO_2 in a few hours. The gas, which is heavier than air, flowed down through two valleys, asphyxiating 1,700 nearby villagers and thousands of cattle. Scientists are studying this tragedy to ensure that no similar man-made event will ever take place. Regulators of storage permits will want assurance that leaks cannot migrate to belowground confined spaces that are vulnerable to sudden release.

Gradual leaks may pose little danger to life, but they could still defeat the climate goals of sequestration. Therefore, researchers are examining the conditions likely to result in slow seepage. Carbon dioxide, which is buoyant in brine, will rise until it hits an impermeable geologic layer (caprock) and can ascend no farther.

Carbon dioxide in a porous formation is like hundreds of helium balloons, and the solid caprock above is like a circus tent. A balloon may escape if the tent has a tear in it or if its surface is tilted to allow a path for the balloon to move sideways and up. Geologists will have to search for faults in the caprock that could allow escape as well as determine the amount of injection pressure that could fracture it. They will also evaluate the very slow horizontal flow of the carbon dioxide outward from the injection locations. Often the sedimentary formations are huge, thin pancakes. If carbon dioxide is injected near the middle of a pancake with a slight tilt, it may not reach the edge for tens of thousands of years. By then, researchers believe, most of the gas will have dissolved in the brine or have been trapped in the pores.

Even if the geology is favorable, using storage formations where there are old wells may be problematic. More than a million wells have been drilled in Texas, for example, and many of them were filled with cement and abandoned. Engineers are worried that CO_2-laden brine, which is acidic, could find its way from an injection well to an abandoned well and thereupon corrode the cement plug and leak to the surface. To find out, some researchers are now exposing cement to brine in the laboratory and sampling old cements from wells. This kind of failure is less likely in carbonate formations than in sandstone ones; the former reduce the destructive potency of the brine.

The world's governments must soon decide how long storage should be maintained. Environmental ethics and traditional economics give different answers. Following a strict environmental ethic that seeks to minimize the impact of today's activities on future generations, authorities might, for instance, refuse to certify a storage project estimated to retain carbon dioxide for only 200 years. Guided instead by traditional economics, they might approve the same project on the grounds that two centuries from now a smarter world will have invented superior carbon disposal technology.

The next few years will be critical for the development of carbon dioxide capture-and-storage methods, as policies evolve that help to make CO_2-emission reduction profitable and as licensing of storage sites gets under way. In conjunction with significant investments in improved energy efficiency, renewable energy sources and, possibly, nuclear energy, commitments to capture and storage can reduce the risks of global warming.

FURTHER READING

CAPTURING GREENHOUSE GASES. Howard Herzog, Baldur Eliasson and Olav Kaarstad in *Scientific American,* Vol. 282, No. 2, pages 72–79; February 2000.

PROSPECTS FOR CO_2 CAPTURE AND STORAGE. International Energy Agency, OECD/IEA, 2004.

STABILIZATION WEDGES: SOLVING THE CLIMATE PROBLEM FOR THE NEXT 50 YEARS WITH CURRENT TECHNOLOGIES. S. Pacala and R. Socolow in *Science,* Vol. 305, pages 968–972; August 13, 2004.

PROSPECTS FOR CARBON CAPTURE AND STORAGE TECHNOLOGIES. Soren Anderson and Richard Newell in *Annual Review of Environment and Resources,* Vol. 29, pages 109–142; 2004.

CARBON DIOXIDE CAPTURE FOR STORAGE IN DEEP GEOLOGICAL FORMATIONS—RESULTS FROM THE CO_2 CAPTURE PROJECT. Two volumes. Edited by David C. Thomas (Vol. 1) and Sally M. Benson (Vol. 2). Elsevier, 2005.

Princeton University Carbon Mitigation Initiative: www.princeton.edu/~cmi

Intergovernmental Panel on Climate Change (IPCC): www.ipcc.ch/index.html (See "Special Report on Carbon Dioxide Capture and Storage," 2005.)

International Energy Agency (IEA) Greenhouse Gas R&D Program: www.ieagreen.org.uk/index.html

Office of Fossil Energy, U.S. Department of Energy: www.fe.doe.gov/programs/sequestration/

CO_2 Capture Project: www.co2captureproject.org

Chaotic Climate

WALLACE S. BROECKER

ORIGINALLY PUBLISHED IN NOVEMBER 1995

The past 10,000 years are anomalous in the history of our planet. This period, during which civilization developed, was marked by weather more consistent and equable than any similar time span of the past 100 millennia. Cores drilled through several parts of the Greenland ice cap show a series of cold snaps and warm spells—each lasting 1,000 years or more—that raised or lowered the average winter temperature in northern Europe by as much as 10 degrees Celsius over the course of as little as a decade. The signs of these sudden changes can be read in the records of atmospheric dust, methane content and precipitation preserved in the annual ice layers.

The last millennium-long cold period, known as the Younger Dryas (after a tundra flower whose habitat expanded significantly), ended about 11,000 years ago. Its marks can be found in North Atlantic marine sediments, Scandinavian and Icelandic glacial moraines, and northern European and maritime Canadian lakes and bogs. New England also cooled significantly.

Further evidence is accumulating that the Younger Dryas's effects were global in scope. The postglacial warming of Antarctica's polar plateau came to a halt for 1,000 years; at the same time, New Zealand's mountain glaciers made a major advance, and the proportions of different species in the plankton population of the South China Sea changed markedly. The atmosphere's methane content dropped by 30 percent. Only pollen records from parts of the U.S. fail to show the period's impact.

THE GREAT CONVEYOR

What lies behind this turbulent history, and could it repeat itself? Although no one knows for sure, there are some very powerful clues. A variety of models suggest that the circulation of heat and salt through the world's oceans can change suddenly, with drastic effects on the global

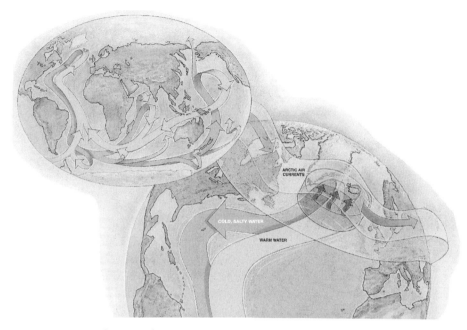

Global conveyor (*dark arrows*) carries cold, salty water, initially formed in the North Atlantic, throughout the world's oceans (*smaller map*). As warm water flows northward to replace it, the resulting transfer of heat has strong climate effects (*larger map*). Northern Europe owes its equable temperatures to the heat that surface water delivers to Arctic air currents (*light arrows*).

climate. Giant, conveyor-like circulation cells span the length of each ocean. In the Atlantic, warm upper waters flow northward, reaching the vicinity of Greenland [*see illustration on these two pages*], where the Arctic air cools them, allowing them to sink and form a current that flows all the way to the Southern Ocean, adjacent to Antarctica. There, warmer and thus less dense than the frigid surface water, the current rises again, is chilled to the freezing point and sinks back into the abyss. Tongues of Antarctic bottom water, the densest in the world, flood northward into the Atlantic, Pacific and Indian oceans, eventually welling up again to repeat the cycle. In the Pacific and Indian oceans, the northward flow of bottom waters is balanced by a southward movement of surface waters. In the Atlantic, this northward counterflow is rapidly entrained into the much stronger southward current of the conveyor.

This so-called deep water forms in the North Atlantic—but not the Pacific—because surface waters in the Atlantic are several percent saltier than those in the Pacific. The locations of large mountain ranges in the Americas, Europe and Africa lead to weather patterns that cause the air

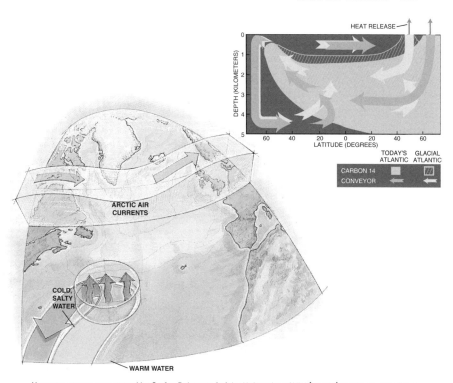

Alternate conveyor proposed by Stefan Rahmstorf of the University of Kiel (*above*) would operate at the latitude of southern Europe and so would not transfer heat effectively to North Atlantic winds. Temperatures in Europe during glacial times, when this conveyor was running, averaged as much as 10 degrees lower than today's. Shallow circulation characterized this alternate conveyor (*top, right*).

leaving the Atlantic basin to be wetter than when it enters; the resulting net loss of water from the surface leads to an excess of salt. Salt makes the upper layers of water denser; as a result, they descend in the North Atlantic and begin a global circulation pattern that effectively redistributes the salt throughout the world's oceans.

The Atlantic's conveyor circulation, which has a flow equal to that of 100 Amazon Rivers, results in an enormous northward transport of heat. The water flowing north is, on average, eight degrees warmer than the cold water flowing south. The transfer of this heat to Arctic air masses over the North Atlantic accounts for the anomalously warm climate enjoyed by Europe.

This pattern, however, is vulnerable to disruption by injections of excess freshwater into the North Atlantic. Precipitation and continental runoff exceed evaporation at high latitudes, and so the salinity of North

Atlantic surface waters depends on the rapidity with which the conveyor sweeps away the excess freshwater delivered by rain and rivers. Any shutdown of the conveyor system would tend to perpetuate itself. Were the conveyor to stop, winter temperatures in the North Atlantic and its surrounding lands would abruptly fall by five or more degrees. Dublin would acquire the climate of Spitsbergen, almost 1,000 kilometers north of the Arctic Circle. Furthermore, the shift would occur in 10 years or less. (Ice cores and other records suggest that the average temperature throughout the North Atlantic basin dropped about seven degrees during ancient cold snaps.)

Ocean modelers have shown that the oceanic conveyor would come back to life, but only after hundreds or thousands of years had passed. Heat mixed down from warm parts of the sea surface and diffusion of salt from the bottom toward the surface would eventually reduce the density of the stagnated deep water, to the point where surface waters from one or the other polar region could once again penetrate into the abyss, reestablishing the circulation of heat and salt. The pattern of this rejuvenated circulation need not be the same as that which existed before the shutdown, however. Instead it would depend on the details of the freshwater runoff patterns for each polar region.

More recently, modeling work by Stefan Rahmstorf of the University of Kiel has suggested that the shutdown of the primary conveyor system may be followed by the formation of an alternate circulation pattern that operates at a shallower depth, with deep water forming north of Bermuda instead of near Greenland. This shift renders the heat released far less effective in warming northern Europe. Rahmstorf's shallow conveyor can be knocked out of action by a pulse of freshwater, just like the primary one, but his model predicts a spontaneous reactivation after only a few decades. It is still not clear, however, how the ocean circulation might switch back from the shallow conveyor to the deeper one that operates today.

Two properties of Rahmstorf's model catch the eye of paleoclimatologists. First, the shallow draft of the alternate conveyor reproduces the ice age distribution of cadmium and carbon isotopes captured in the shells of tiny bottom-dwelling creatures called benthic foraminifera. Today the waters of the North Atlantic conveyor are poor in cadmium and rich in carbon 13, whereas deep waters in the rest of the ocean are rich in cadmium and poor in carbon 13.

This contrast reflects the fact that respiration by aquatic organisms depletes carbon 13 and enhances the concentration of cadmium (and other constituents whose history is not recorded in benthic shells). During cold episodes, cadmium levels dropped in the mid-depth Atlantic waters and rose dramatically in the bottom waters; the ratio of carbon 13 to carbon 12 displayed the opposite pattern—consistent with Rahmstorf's conclusion that the conveyor operated at a shallower depth and bypassed the bottom-most water.

Second, the alternate conveyor maintains the movement of radiocarbon into the deep sea. If this transfer had ceased, radiochemical-dating methods based on carbon 14 decay would show huge distortions; in fact, the radiocarbon clock has been calibrated by other means and found to be imperfect but still basically valid.

Only about a quarter of the world's carbon currently resides in the upper ocean and atmosphere. The remainder is in the abyss. The distribution of radioactive carbon 14, which is formed in the atmosphere by cosmic rays, depends on the rate of oceanic circulation. In today's ocean, most of the radiocarbon reaching the deep sea does so via the Atlantic's conveyor circulation. During their traverse up the Atlantic, waters in the conveyor's warm upper limb are recharged with radiocarbon by absorption from the air. The conveyor then carries this radiocarbon down to the ocean depths. Although the deep water resurfaces briefly in the region around the Antarctic continent, little radiocarbon finds its way into solution there.

This state of affairs implies that even a slowdown of the conveyor would have a significant effect on the abundance of carbon 14 in both the atmosphere and the ocean. The ratio of carbon 14 to stable carbon 12 in the deep ocean is at present approximately 12 percent lower than the average for the upper ocean and the atmosphere because of the radioactive decay that takes place while the deep water is circulating. Meanwhile cosmic rays replenish 1 percent of the world's radiocarbon inventory every 82 years. As a result, if exchanges between the upper and deep ocean were to cease, the carbon 14 ratio in the upper ocean and the atmosphere would rise at the rate of 5 percent every century because carbon 14 was being added but not swept down into the deep sea. After a millennium of isolation, the atmosphere's carbon 14 ratio would rise by a third of its original value.

Such an occurrence would lead to a radical disturbance of the radiocarbon-dating record. Paleontologists determine the age of organic materials by measuring their residual carbon 14 content. The amount incorporated into a plant's structure while it is alive depends on the proportion of radiocarbon in the atmosphere (or ocean) at the time; the less carbon 14 that remains, the older a specimen must be. Plants that grew during a conveyor shutdown would incorporate the extra carbon 14 and appear younger than their true age. Then, when the conveyor started up again and brought atmospheric carbon 14 back down near its current level, the anomaly would disappear. Thus, plants from the cold times would appear—according to carbon dating—to be contemporary with warm-weather specimens that lived more than 1,000 years later.

Although the amount of carbon 14 in the atmosphere has varied somewhat over time, sequences of radiocarbon dates from marine sediments likely to have accumulated at a nearly uniform rate clearly demonstrate that no such sudden shock took place at any time during the past 20,000 years. Indeed, measurements on corals whose absolute ages have been established by uranium-thorium dating imply that during the end of the last ice age, when the conveyor should have been starting up again and drawing carbon 14 out of the atmosphere, the radiocarbon content of the atmosphere increased.

This record seems to be telling us that any conveyor shutdowns must have been brief—a century or less—and that they must have been matched by intervening intervals of rapid mixing. In particular, the Younger Dryas was apparently a time when overall ocean circulation increased rather than decreased, as would be expected if the cold snap were caused by a complete halt of the Atlantic conveyor. If the conveyor did shut down, some other method of transporting carbon 14 to the deep sea must have been in operation.

A FLEET OF ICEBERGS

Assuming that changes in the conveyor mechanism did drive the abrupt changes found in the Greenland ice cores and other climate records, what might supply the excess freshwater needed to shut down transport of water into the abyss? The polar ice caps are the obvious sources for the jolts of freshwater needed to upset ocean circulation. Moreover, sudden changes appear to be confined to times when large ice sheets covered Canada and Scandinavia. Since the ice ages ended, global climate has remained locked in its present mode.

There is evidence of at least eight invasions of freshwater into the North Atlantic: seven armadas of icebergs released from the eastern margin of the Hudson Bay ice cap and a flood of meltwater from a huge lake that marked the southern margin of the ice sheet during glacial retreat. In the early 1980s, while he was a graduate student at the University of Göttingen, Hartmut Heinrich discovered a curious set of layers in the sediments of the North Atlantic. The layers stretch from the Labrador Sea to the British Isles, and their characteristics are most plausibly explained by the melting of enormous numbers of icebergs launched from Canada.

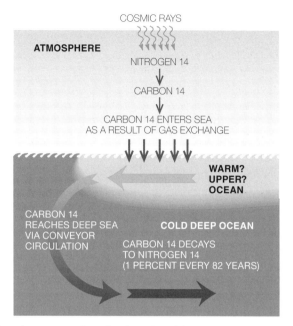

COSMIC RAYS

ATMOSPHERE

NITROGEN 14

CARBON 14

CARBON 14 ENTERS SEA
AS A RESULT OF GAS EXCHANGE

**WARM?
UPPER?
OCEAN**

CARBON 14
REACHES DEEP SEA
VIA CONVEYOR
CIRCULATION

COLD DEEP OCEAN

CARBON 14 DECAYS
TO NITROGEN 14
(1 PERCENT EVERY 82 YEARS)

Deepwater formation process carries radioactive carbon 14, formed by cosmic rays, out of the atmosphere and upper ocean and into the abyss. Radiocarbon dating indirectly measures the state of the oceanic conveyor because any prolonged shutdown causes a buildup of carbon 14 in the atmosphere and scrambles the apparent ages of organic remains.

The debris dropped from this flotilla thins eastward from half a meter in the Labrador Sea to a few centimeters in the eastern Atlantic. Rock fragments characteristic of the sedimentary limestones and igneous bedrock from Hudson Bay and the surrounding area constitute most of the larger particles of the sediments. Shells of foraminifera are found only rarely in these layers, suggesting an ocean choked with sea ice; the low ratio of oxygen 18 to oxygen 16 in those shells that do appear provides an unambiguous marker that the animals lived in water much less salty than usual. (Rain and snow at high latitudes is depleted in oxygen 18 because the "heavy" water containing it condenses out of the atmosphere preferentially as air masses cool.)

The eighth freshwater pulse came from Lake Agassiz, a very large lake trapped in the topographic depression created by the weight of the retreating ice cap. Initially, the water from the lake spilled over a rock sill into the Mississippi River watershed and thence the Gulf of Mexico. About 12,000 years ago the retreat of the ice front opened a channel to the east, triggering a catastrophic drop in lake level. The water released during this breakthrough flooded across southern Canada into the valley now

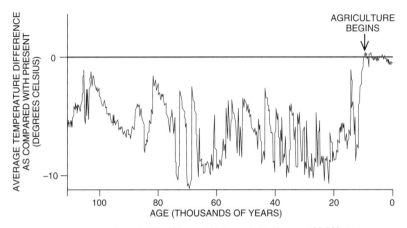

Ice core data show the variability of the earth's climate during the past 100,000 years.

occupied by the St. Lawrence River and discharged directly into the region where deep waters now form.

The connection between these events and local climate changes is clear. Four occurred at times corresponding to significant changes in the climate of the North Atlantic basin. One of Heinrich's layers marks the end of the second-to-last major glacial cycle, and another that of the most recent cycle. A third layer appears to match the onset of glacial conditions in the North Atlantic, and the catastrophic release of water from Lake Agassiz coincided with the onset of the Younger Dryas. Each of the four remaining pulses caps a climate subcycle. Gerard C. Bond of the Lamont-Doherty Earth Observatory of Columbia University correlated Heinrich layers with the Greenland ice core record and found that the millennia-long cold events come in groups characterized by progressively more severe cold snaps, culminating with a Heinrich event that is followed in turn by a significant warming that begins a new cycle.

The climate shifts of the Younger Dryas period were felt around the world. Is the same true of the 15 or so similar events that appear earlier in the ice core record? So far only two pieces of evidence point in that direction, but they are convincing ones. First, Jerome A. Chappellaz of the Laboratory of Glaciology and Geophysics of the Environment near Grenoble has analyzed air trapped in Greenland ice cores and found that cold periods were accompanied by drops in the atmosphere's methane content. Methane is produced mostly in swamps and bogs. Because those of the northern temperate region were either frozen or buried underneath ice during glacial times, methane present in the atmosphere must have come from the tropics. The fluctuations in the methane record imply that the tropics dried out during each of the northern cold intervals.

The second clue is an as yet unpublished study by James P. Kennett and Richard J. Behl of the University of California at Santa Barbara of a sediment core recovered from 500 meters below sea level in the Santa Barbara basin. The two found that bands of undisturbed sediment with clear annual layers alternated with sections more or less disturbed by burrowing worms. The presence of worms implies that the bottom water in the area contained significant amounts of oxygen, enough to support life; such periods display an uncanny correlation with cold spells in Greenland, implying that changes in ocean circulation reached around the globe.

More surprising is the finding that Heinrich events also appear to have had a worldwide imprint. Eric Grimm of the Illinois State Museum and his colleagues sampled pollen in the sediments of Lake Tulane in Florida and found one prominent peak in the ratio of pine to oak for each Heinrich event. Pine trees prosper in relatively wet climates, whereas oaks prefer dryer ones. Although the exact relation between pine-rich intervals and Heinrich events awaits more accurate radiocarbon dating, the Lake Tulane record suggests one wet interval per cycle. George H. Denton of the University of Maine and his co-workers found an even more distant connection: each of the four Heinrich events falling within the range of radiocarbon dating matches a sharp maximum in the extent of Andean mountain glaciers.

The finding that the massive calving of Canadian glaciers caused global impacts creates a paradox. Atmospheric models indicate clearly that climate shifts related to changes in the amount of heat delivered to the atmosphere over the North Atlantic would be limited to the surrounding regions. The evidence that has been found, however, demands a mechanism for extending these effects to the tropics, the southern temperate region and even the Antarctic.

The symmetrical distribution of these climate changes around the equator points to the tropics. Changes in the dynamics of the tropical atmosphere could easily have a far-reaching effect. The towering convection cells that form in the tropical atmosphere where the trade winds meet feed the atmosphere with its dominant greenhouse gas: water vapor. Although the link between ocean circulation and tropical convection is tenuous, it seems plausible that changed circulation patterns might alter the amount of cold water upwelling to the surface along the equator in the Pacific. This upwelling is an important part of the region's heat budget and thus its overall climate. Reduction of the equatorial upwelling, as occurs now during so-called El Niño periods, can cause droughts in some areas and floods in others.

GLOBAL SHIFTS

Support for such a scenario comes not only from Chappellaz's data showing drying in the tropics but also from the moisture histories of Nevada, New Mexico, Texas, Florida and Virginia. The most dramatic evidence comes from the Great Basin area of the western U.S.: immediately after the last Heinrich event about 14,000 years ago, Lake Lahontan in Nevada achieved its greatest size, an order of magnitude larger than today's remnant. Supporting such a large body of water requires immense amounts of precipitation, of the magnitude experienced during the record El Niño winter of 1982–1983. One way of thinking about the impact of these earlier occurrences, then, is as changes in the pattern of ocean circulation that led to El Niños lasting 1,000 years.

More recent findings, from Lonnie G. Thompson of Ohio State University, reinforce the evidence that tropical weather was extremely different during glacial times. Ancient ice cores from 6,000 meters up in the tropical Andes contain 200 times as much fine dust as more recent samples— dust probably carried by winds blowing up from an arid Amazonia. The older ice is also depleted in oxygen 18 as compared with ice formed more recently than 10,000 years ago, implying a temperature about 10 degrees lower than today. Taken with the observation that the Andean snow line reached down a full 1,000 meters closer to sea level during the ice ages, these data suggest that the tropics of glacial times were both colder and drier.

The conclusion that the earth's climatic system has occasionally jumped from one mode of operation to another is rock solid. Unfortunately, researchers have yet to pin down the cause of these abrupt shifts. Although large-scale reorganizations of the ocean's circulation seem the most likely candidate, it is possible that atmospheric triggers may be discovered as well.

A FRAGILE BALANCE

This situation leaves us in limbo with regard to climatic prediction. Might the current buildup of greenhouse gases set in motion yet another reorganization of the deepwater conveyor and the weather patterns that depend on it? On the one hand, the paleographic record suggests that jumps have been confined to times when the North Atlantic was surrounded by huge ice sheets, a situation that is now further from the case than ever. On the other hand, the greenhouse nudge promises to be far larger than any other forcing experienced during an interglacial interval, and there is no certainty that the system will remain locked in its present relatively benign mode.

A conveyor shutdown or comparable drastic change is unlikely, but were it to occur, the impact would be catastrophic. The likelihood of such an event will be highest between 50 and 150 years from now, at a time when the world will be bulging with people threatened by hunger and disease and struggling to maintain wildlife under escalating environmental pressure. It behooves us to take this possibility seriously. We should spare no effort in the attempt to understand better the chaotic behavior of the global climatic system.

FURTHER READING

WHAT DRIVES GLACIAL CYCLES? Wallace S. Broecker and George H. Denton in *Scientific American*, Vol. 262, No. 1, pages 48–56; January 1990.

EVIDENCE FOR GENERAL INSTABILITY OF PAST CLIMATE FROM A 250-KYR ICE-CORE RECORD. W. Dansgaard et al. in *Nature*, Vol. 364, pages 218–220; July 15, 1993.

A LOW-ORDER MODEL OF THE HEINRICH EVENT CYCLE. D. R. MacAyeal in *Paleoceanography*, Vol. 8. No. 6, pages 767–773; December 1993.

SYNCHRONOUS CHANGES IN ATMOSPHERIC CH$_4$ AND GREENLAND CLIMATE BETWEEN 40 AND 8 KYR BP. J. Chappellaz, T. Blunier, D. Raynaud, J. M. Barnola, J. Schwander and B. Stauffer in *Nature*, Vol. 366, pages 443–445; December 2, 1993.

Defusing the Global Warming Time Bomb

JAMES HANSEN

ORIGINALLY PUBLISHED IN MARCH 2004

Aparadox in the notion of human-made global warming became strikingly apparent to me one summer afternoon in 1976 on Jones Beach, Long Island. Arriving at midday, my wife, son and I found a spot near the water to avoid the scorching hot sand. As the sun sank in the late afternoon, a brisk wind from the ocean whipped up whitecaps. My son and I had goose bumps as we ran along the foamy shoreline and watched the churning waves.

That same summer Andy Lacis and I, along with other colleagues at the NASA Goddard Institute for Space Studies, had estimated the effects of greenhouse gases on climate. It was well known by then that human-made greenhouse gases, especially carbon dioxide and chlorofluorocarbons (CFCs), were accumulating in the atmosphere. These gases are a climate "forcing," a perturbation imposed on the energy budget of the planet. Like a blanket, they absorb infrared (heat) radiation that would otherwise escape from the earth's surface and atmosphere to space.

Our group had calculated that these human-made gases were heating the earth's surface at a rate of almost two watts per square meter. A miniature Christmas tree bulb dissipates about one watt, mostly in the form of heat. So it was as if humans had placed two of these tiny bulbs over every square meter of the earth's surface, burning night and day.

The paradox that this result presented was the contrast between the awesome forces of nature and the tiny lightbulbs. Surely their feeble heating could not command the wind and waves or smooth our goose bumps. Even their imperceptible heating of the ocean surface must be quickly dissipated to great depths, so it must take many years, perhaps centuries, for the ultimate surface warming to be achieved.

This seeming paradox has now been largely resolved through study of the history of the earth's climate, which reveals that small forces, maintained long enough, can cause large climate change. And, consistent with the historical evidence, the earth has begun to warm in recent decades at a rate predicted by climate models that take account of the atmospheric accumulation of human-made greenhouse gases. The warming is having noticeable impacts as glaciers are retreating worldwide, Arctic sea ice has thinned, and spring comes about one week earlier than when I grew up in the 1950s.

Yet many issues remain unresolved. How much will climate change in coming decades? What will be the practical consequences? What, if anything, should we do about it? The debate over these questions is highly charged because of the inherent economic stakes.

Objective analysis of global warming requires quantitative knowledge of three issues: the sensitivity of the climate system to forcings, the forcings that humans are introducing, and the time required for climate to respond. All these issues can be studied with global climate models, which are numerical simulations on computers. But our most accurate knowledge about climate sensitivity, at least so far, is based on empirical data from the earth's history.

THE LESSONS OF HISTORY

Over the past few million years the earth's climate has swung repeatedly between ice ages and warm interglacial periods. A 400,000-year record of temperature is preserved in the Antarctic ice sheet, which, except for coastal fringes, escaped melting even in the warmest interglacial periods. This record [*see box on opposite page*] suggests that the present interglacial period (the Holocene), now about 12,000 years old, is already long of tooth.

The natural millennial climate swings are associated with slow variations of the earth's orbit induced by the gravity of other planets, mainly Jupiter and Saturn (because they are so heavy) and Venus (because it comes so close). These perturbations hardly affect the annual mean solar energy striking the earth, but they alter the geographical and seasonal distribution of incoming solar energy, or insolation, as much as 20 percent. The insolation changes, over long periods, affect the building and melting of ice sheets.

Insolation and climate changes also affect uptake and release of carbon dioxide and methane by plants, soil and the ocean. Climatologists are still developing a quantitative understanding of the mechanisms by which the ocean and land release carbon dioxide and methane as the earth warms, but the paleoclimate data are already a gold mine of information. The most critical insight that the ice age climate swings provide is an empirical measure of climate sensitivity.

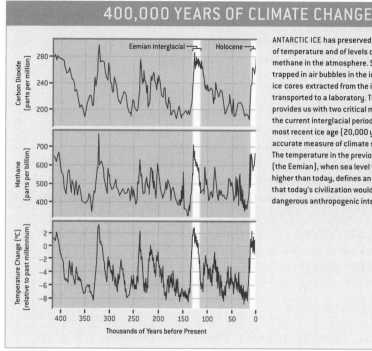

400,000 YEARS OF CLIMATE CHANGE

ANTARCTIC ICE has preserved a 400,000-year record of temperature and of levels of carbon dioxide and methane in the atmosphere. Scientists study gases trapped in air bubbles in the ice—generally using ice cores extracted from the ice sheet and transported to a laboratory. The historical record provides us with two critical measures: Comparison of the current interglacial period (the Holocene) with the most recent ice age (20,000 years ago) gives an accurate measure of climate sensitivity to forcings. The temperature in the previous interglacial period (the Eemian), when sea level was several meters higher than today, defines an estimate of the warming that today's civilization would consider to be dangerous anthropogenic interference with climate.

The composition of the ice age atmosphere is known precisely from air bubbles trapped as the Antarctic and Greenland ice sheets and numerous mountain glaciers built up from annual snowfall. Furthermore, the geographical distributions of the ice sheets, vegetation cover and coastlines during the ice age are well mapped. From these data we know that the change of climate forcing between the ice age and today was about 6.5 watts per square meter. This forcing maintains a global temperature change of 5 degrees Celsius (9 degrees Fahrenheit), implying a climate sensitivity of 0.75 ± 0.25 degrees C per watt per square meter. Climate models yield a similar climate sensitivity. The empirical result is more precise and reliable, however, because it includes all the processes operating in the real world, even those we have not yet been smart enough to include in the models.

The paleodata provide another important insight. Changes of the earth's orbit instigate climate change, but they operate by altering atmosphere and surface properties and thus the planetary energy balance. These atmosphere and surface properties are now influenced more by humans than by our planet's orbital variations.

CLIMATE-FORCING AGENTS TODAY

The largest change of climate forcings in recent centuries is caused by human-made greenhouse gases. Greenhouse gases in the atmosphere absorb heat radiation rather than letting it escape into space. In effect, they make the proverbial blanket thicker, returning more heat toward the ground rather than letting it escape to space. The earth then is radiating less energy to space than it absorbs from the sun. This temporary planetary energy imbalance results in the earth's gradual warming.

Because of the large capacity of the oceans to absorb heat, it takes the earth about a century to approach a new balance—that is, for it to once again receive the same amount of energy from the sun that it radiates to space. And of course the balance is reset at a higher temperature. In the meantime, before it achieves this equilibrium, more forcings may be added.

The single most important human-made greenhouse gas is carbon dioxide, which comes mainly from burning fossil fuels (coal, oil and gas). Yet the combined effect of the other human-made gases is comparable. These other gases, especially tropospheric ozone and its precursors, including methane, are ingredients in smog that damage human health and agricultural productivity.

Aerosols (fine particles in the air) are the other main human-made climate forcing. Their effect is more complex. Some "white" aerosols, such as sulfates arising from sulfur in fossil fuels, are highly reflective and thus reduce solar heating of the earth; however, black carbon (soot), a product of incomplete combustion of fossil fuels, biofuels and outdoor biomass burning, absorbs sunlight and thus heats the atmosphere. This aerosol direct climate forcing is uncertain by at least 50 percent, in part because aerosol amounts are not well measured and in part because of their complexity.

Aerosols also cause an indirect climate forcing by altering the properties of clouds. The resulting brighter, longer-lived clouds reduce the amount of sunlight absorbed by the earth, so the indirect effect of aerosols is a negative forcing that causes cooling.

Other human-made climate forcings include replacement of forests by cropland. Forests are dark even with snow on the ground, so their removal reduces solar heating.

Natural forcings, such as volcanic eruptions and fluctuations of the sun's brightness, probably have little trend on a timescale of 1,000 years. But evidence of a small solar brightening over the past 150 years implies a climate forcing of a few tenths of a watt per square meter.

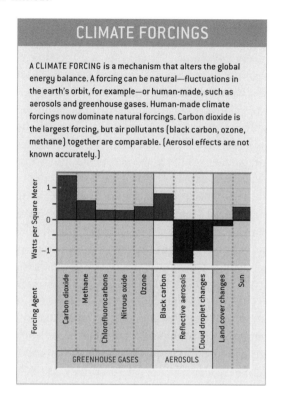

CLIMATE FORCINGS

A CLIMATE FORCING is a mechanism that alters the global energy balance. A forcing can be natural—fluctuations in the earth's orbit, for example—or human-made, such as aerosols and greenhouse gases. Human-made climate forcings now dominate natural forcings. Carbon dioxide is the largest forcing, but air pollutants (black carbon, ozone, methane) together are comparable. (Aerosol effects are not known accurately.)

The net value of the forcings added since 1850 is 1.6 ± 1.0 watts per square meter. Despite the large uncertainties, there is evidence that this estimated net forcing is approximately correct. One piece of evidence is the close agreement of observed global temperature during the past several decades with climate models driven by these forcings. More fundamentally, the observed heat gain by the world ocean in the past 50 years is consistent with the estimated net climate forcing.

GLOBAL WARMING

Global average surface temperature has increased about 0.75 degree C during the period of extensive instrumental measurements, which began in the late 1800s. Most of the warming, about 0.5 degree C, occurred after 1950. The causes of observed warming can be investigated best for the past 50 years, because most climate forcings were observed then, especially since satellite measurements of the sun, stratospheric aerosols and ozone began in the 1970s. Furthermore, 70 percent of the anthropogenic increase of greenhouse gases occurred after 1950.

The most important quantity is the planetary energy imbalance [*see box on page 302*]. This imbalance is a consequence of the long time that it takes the ocean to warm. We conclude that the earth is now out of balance by something between 0.5 and one watt per square meter—that much more solar radiation is being absorbed by the earth than is being emitted as heat to space. Even if atmospheric composition does not change further, the earth's surface will therefore eventually warm another 0.4 to 0.7 degree C.

Most of the energy imbalance has been heat going into the ocean. Sydney Levitus of the National Oceanic and Atmospheric Administration has analyzed ocean temperature changes of the past 50 years, finding that the world ocean heat content increased about 10 watt-years per square meter in the past 50 years. He also finds that the rate of ocean heat storage in recent years is consistent with our estimate that the earth is now out of energy balance by 0.5 to one watt per square meter. Note that the amount of heat required to melt enough ice to raise sea level one meter is about 12 watt-years (averaged over the planet), energy that could be accumulated in 12 years if the planet is out of balance by one watt per square meter.

The agreement with observations, for both the modeled temperature change and ocean heat storage, leaves no doubt that observed global climate change is being driven by natural and anthropogenic forcings. The current rate of ocean heat storage is a critical planetary metric: it not only determines the amount of additional global warming already in the pipeline, but it also equals the reduction in climate forcings needed to stabilize the earth's present climate.

THE TIME BOMB

The goal of the United Nations Framework Convention on Climate Change, produced in Rio de Janeiro in 1989, is to stabilize atmospheric composition to "prevent dangerous anthropogenic interference with the climate system" and to achieve that goal in ways that do not disrupt the global economy. Defining the level of warming that constitutes "dangerous anthropogenic interference" is thus a crucial but difficult part of the problem.

The U.N. established an Intergovernmental Panel on Climate Change (IPCC) with responsibility for analysis of global warming. The IPCC has defined climate-forcing scenarios, used these for simulations of 21st-century climate, and estimated the impact of temperature and precipitation changes on agriculture, natural ecosystems, wildlife and other matters. The IPCC estimates sea-level change as large as several tens of centimeters in 100 years, if global warming reaches several degrees Celsius. The group's calculated sea-level change is due mainly to thermal expansion of ocean water, with little change in ice-sheet volume.

These moderate climate effects, even with rapidly increasing greenhouse gases, leave the impression that we are not close to dangerous anthropogenic interference. I will argue, however, that we are much closer than is generally realized, and thus the emphasis should be on mitigating the changes rather than just adapting to them.

The dominant issue in global warming, in my opinion, is sea-level change and the question of how fast ice sheets can disintegrate. A large portion of the world's people live within a few meters of sea level, with trillions of dollars of infrastructure. The need to preserve global coastlines sets a low ceiling on the level of global warming that would constitute dangerous anthropogenic interference.

The history of the earth and the present human-made planetary energy imbalance together paint a disturbing picture about prospects for sea-level change. Data from the Antarctic temperature record show that the warming of the past 50 years has taken global temperature back to approximately the peak of the current interglacial (the Holocene). There is some additional warming in the pipeline that will take us about halfway to the highest global temperature level of the previous interglacial (the Eemian), which was warmer than the Holocene, with sea level estimated to have been five to six meters higher. One additional watt per square meter of forcing, over and above that today, will take global temperature approximately to the maximum level of the Eemian.

The main issue is: How fast will ice sheets respond to global warming? The IPCC calculates only a slight change in the ice sheets in 100 years; however, the IPCC calculations include only the gradual effects of changes in snowfall, evaporation and melting. In the real world, ice-sheet disintegration is driven by highly nonlinear processes and feedbacks. The peak rate of deglaciation following the last ice age was a sustained rate of melting of more than 14,000 cubic kilometers a year—about one meter of sea-level rise every 20 years, which was maintained for several centuries. This period of most rapid melt coincided, as well as can be measured, with the time of most rapid warming.

Given the present unusual global warming rate on an already warm planet, we can anticipate that areas with summer melt and rain will expand over larger areas of Greenland and fringes of Antarctica. Rising sea level itself tends to lift marine ice shelves that buttress land ice, unhinging them from anchor points. As ice shelves break up, this accelerates movement of land ice to the ocean. Although building of glaciers is slow, once an ice sheet begins to collapse, its demise can be spectacularly rapid.

The human-induced planetary energy imbalance provides an ample supply of energy for melting ice. Furthermore, this energy source is supplemented by increased absorption of sunlight by ice sheets darkened by black-carbon aerosols, and the positive feedback process as meltwater darkens the ice surface.

These considerations do not mean that we should expect large sea-level change in the next few years. Preconditioning of ice sheets for accelerated breakup may require a long time, perhaps many centuries. (The satellite ICESat, recently launched by NASA, may be able to detect early signs of accelerating ice-sheet breakup.) Yet I suspect that significant sea-level rise could begin much sooner if the planetary energy imbalance continues to increase. It seems clear that global warming beyond some limit will make a large sea-level change inevitable for future generations. And once large-scale ice-sheet breakup is under way, it will be impractical to stop. Dikes may protect limited regions, such as Manhattan and the Netherlands, but most of the global coastlines will be inundated.

I argue that the level of dangerous anthropogenic influence is likely to be set by the global temperature and planetary radiation imbalance at which substantial deglaciation becomes practically impossible to avoid. Based on the paleoclimate evidence, I suggest that the highest prudent level of additional global warming is not more than about one degree C. This means that additional climate forcing should not exceed about one watt per square meter.

CLIMATE-FORCING SCENARIOS

The IPCC defines many climate-forcing scenarios for the 21st century based on multifarious "story lines" for population growth, economic development and energy sources. It estimates that added climate forcing in the next 50 years is one to three watts per square meter for carbon dioxide and two to four watts per square meter with other gases and aerosols included. Even the IPCC's minimum added forcing would cause dangerous anthropogenic interference with the climate system based on our criterion.

The IPCC scenarios may be unduly pessimistic, however. First, they ignore changes in emissions, some already under way, because of concerns

EARTH'S ENERGY IMBALANCE

THE EARTH'S ENERGY is balanced when the outgoing heat from the earth equals the incoming energy from the sun. At present the energy budget is not balanced (*diagram* and *table*). Human-made aerosols have increased reflection of sunlight by the earth, but this reflection is more than offset by the trapping of heat radiation by greenhouse gases. The excess energy—about one watt per square meter—warms the ocean and melts ice. The simulated planetary energy imbalance (*graph*) is confirmed by measurements of heat stored in the oceans. The planetary energy imbalance is a critical metric, in that it measures the net climate forcing and foretells future global warming already in the pipeline.

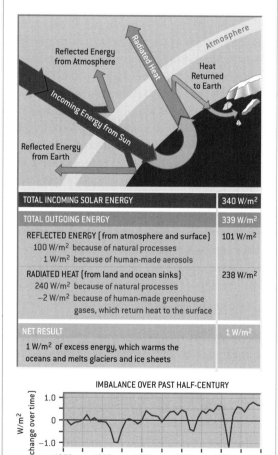

TOTAL INCOMING SOLAR ENERGY	340 W/m^2
TOTAL OUTGOING ENERGY	339 W/m^2

REFLECTED ENERGY (from atmosphere and surface)
    100 W/m^2 because of natural processes
    1 W/m^2 because of human-made aerosols
	101 W/m^2

RADIATED HEAT (from land and ocean sinks)
    240 W/m^2 because of natural processes
    -2 W/m^2 because of human-made greenhouse
        gases, which return heat to the surface
	238 W/m^2

NET RESULT	1 W/m^2
1 W/m^2 of excess energy, which warms the oceans and melts glaciers and ice sheets	

about global warming. Second, they assume that true air pollution will continue to get worse, with ozone, methane and black carbon all greater in 2050 than in 2000. Third, they give short shrift to technology advances that can reduce emissions in the next 50 years.

An alternative way to define scenarios is to examine current trends of climate-forcing agents, to ask why they are changing as observed, and to try to understand whether reasonable actions could encourage further changes in the growth rates.

The growth rate of the greenhouse-gas climate forcing peaked in the early 1980s at almost 0.5 watt per square meter per decade but declined by the 1990s to about 0.3 watt per square meter per decade. The primary reason for the decline was reduced emissions of chlorofluorocarbons, whose production was phased out because of their destructive effect on stratospheric ozone.

The two most important greenhouse gases, with chlorofluorocarbons on the decline, are carbon dioxide and methane. The growth rate of carbon dioxide surged after World War II, flattened out from the mid-1970s to the mid-1990s, and rose moderately in recent years to the current growth rate of about two parts per million per year. The methane growth rate has declined dramatically in the past 20 years, by at least two thirds.

These growth rates are related to the rate of global fossil-fuel use. Fossil-fuel emissions increased by more than 4 percent a year from the end of World War II until 1975 but subsequently by only about 1 percent a year. The change in fossil-fuel growth rate occurred after the oil embargo and price increases of the 1970s, with subsequent emphasis on energy efficiency. Methane growth has also been affected by other factors, including changes in rice farming and increased efforts to capture methane at landfills and in mining operations.

If recent growth rates of these greenhouse gases continued, the added climate forcing in the next 50 years would be about 1.5 watts per square meter. To this must be added the change caused by other forcings, such as atmospheric ozone and aerosols. These forcings are not well monitored globally, but it is known that they are increasing in some countries while decreasing in others. Their net effect should be small, but it could add as much as 0.5 watt per square meter. Thus, if there is no slowing of emission rates, the human-made climate forcing could increase by two watts per square meter in the next 50 years.

This "current trends" growth rate of climate forcings is at the low end of the IPCC range of two to four watts per square meter. The IPCC four watts per square meter scenario requires 4 percent a year exponential growth of carbon dioxide emissions maintained for 50 years and large growth of air pollution; it is implausible.

Nevertheless, the "current trends" scenario is larger than the one watt per square meter level that I suggested as our current best estimate for the level of dangerous anthropogenic influence. This raises the question of whether there is a feasible scenario with still lower climate forcing.

A BRIGHTER FUTURE

I have developed a specific alternative scenario that keeps added climate forcing in the next 50 years at about one watt per square meter. It has two components: first, halt or reverse growth of air pollutants, specifically soot, atmospheric ozone and methane; second, keep average fossil-fuel carbon dioxide emissions in the next 50 years about the same as today. The carbon dioxide and non-carbon dioxide portions of the scenario are equally important. I argue that they are feasible and at the same time protect human health and increase agricultural productivity.

In addressing air pollution, we should emphasize the constituents that contribute most to global warming. Methane offers a great opportunity. If human sources of methane are reduced, it may even be possible to get the atmospheric methane amount to decline, thus providing a cooling that would partially offset the carbon dioxide increase. Reductions of black-carbon aerosols would help counter the warming effect of reductions in sulfate aerosols. Atmospheric ozone precursors, besides methane, especially nitrogen oxides and volatile organic compounds, must be reduced to decrease low-level atmospheric ozone, the prime component of smog.

Actions needed to reduce methane, such as methane capture at landfills and at waste management facilities and during the mining of fossil fuels, have economic benefits that partially offset the costs. In some cases, methane's value as a fuel entirely pays for the cost of capture. Reducing black carbon would also have economic benefits, both in the decreased loss of life and work-years (minuscule soot particles carry toxic organic compounds and metals deep into lungs) and in increased agricultural productivity in certain parts of the world. Prime sources of black carbon are diesel fuels and biofuels (wood and cow dung, for example). These sources need to be dealt with for health reasons. Diesel could be burned more cleanly with improved technologies; however, there may be even better solutions, such as hydrogen fuel, which would eliminate ozone precursors as well as soot.

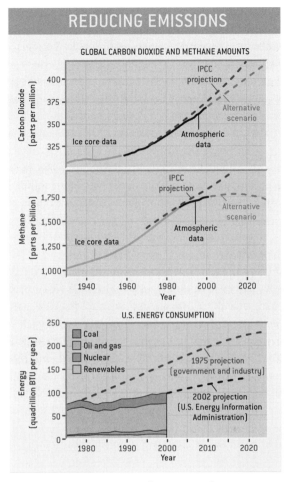

REDUCING EMISSIONS

GLOBAL CARBON DIOXIDE AND METHANE AMOUNTS

Observed amounts of carbon dioxide and methane [*top two graphs*] fall below IPCC estimates, which have proved consistently pessimistic. Although the author's alternative scenario agrees better with observations, continuation on that path requires a gradual slowdown in carbon dioxide and methane emissions. Improvements in energy efficiency [*bottom graph*] have allowed energy use in the U.S. to fall below projections in recent decades, but more rapid efficiency gains are needed to achieve the carbon dioxide emissions of the alternative scenario, unless nuclear power and renewable energies grow substantially.

Improved energy efficiency and increased use of renewable energies might level carbon dioxide emissions in the near term. Long-term reduction of carbon dioxide emissions is a greater challenge, as energy use will continue to rise. Progress is needed across the board: continued efficiency improvements, more renewable energy, and new technologies that produce little or no carbon dioxide or that capture and sequester it. Next-generation nuclear power, if acceptable to the public, could be an important contributor. There may be new technologies before 2050 that we have not imagined.

Observed global carbon dioxide and methane trends [*see box on opposite page*] for the past several years show that the real world is falling below all IPCC scenarios. It remains to be proved whether the smaller observed growth rates are a fluke, soon to return to IPCC rates, or are a meaningful difference. In contrast, the projections of my alternative scenario and the observed growth rates are in agreement. This is not surprising, because that scenario was defined with observations in mind. And in the three years since the alternative scenario was defined, observations have continued on that path. I am not suggesting, however, that the alternative scenario can be achieved without concerted efforts to reduce anthropogenic climate forcings.

How can I be optimistic if climate is closer to the level of dangerous anthropogenic interference than has been realized? If we compare the situation today with that 10 to 15 years ago, we note that the main elements required to halt climate change have come into being with remarkable rapidity. I realize that it will not be easy to stabilize greenhouse-gas concentrations, but I am optimistic because I expect that empirical evidence for climate change and its impacts will continue to accumulate and that this will influence the public, public-interest groups, industry and governments at various levels. The question is: Will we act soon enough?

Meltdown in the North

MATTHEW STURM, DONALD K. PEROVICH AND MARK C. SERREZE

ORIGINALLY PUBLISHED IN OCTOBER 2003

Snow crystals sting my face and coat my beard and the ruff of my parka. As the wind rises, it becomes difficult to see my five companions through the blowing snow. We are 500 miles into a 750-mile snow-mobile trip across Arctic Alaska. We have come, in the late winter of 2002, to measure the thickness of the snow cover and estimate its insulating capacity, an important factor in maintaining the thermal balance of the permafrost. I have called a momentary halt to decide what to do. The rising wind, combined with −30 degree Fahrenheit temperatures, makes it clear we need to find shelter, and fast. I put my face against the hood of my nearest companion and shout: "Make sure everyone stays close together. We have to get off this exposed ridge."

At the time, the irony that we might freeze to death while looking for evidence of global warming was lost on me, but later, snug in our tents, I began to laugh at how incongruous that would have been.
—Matthew Sturm

The list is impressively long: The warmest air temperatures in four centuries, a shrinking sea-ice cover, a record amount of melting on the Greenland Ice Sheet, Alaskan glaciers retreating at unprecedented rates. Add to this the increasing discharge from Russian rivers, an Arctic growing season that has lengthened by several days per decade, and permafrost that has started to thaw. Taken together, these observations announce in a way no single measurement could that the Arctic is undergoing a profound transformation. Its full extent has come to light only in the past decade, after scientists in different disciplines began comparing their findings. Now many of those scientists are collaborating, trying to understand the ramifications of the changes and to predict what lies ahead for the Arctic and the rest of the globe.

What they learn will have planetwide importance because the Arctic exerts an outsize degree of control on the climate. Much as a spillway in a dam controls the level of a reservoir, the polar regions control the earth's heat balance. Because more solar energy is absorbed in the tropics than at the poles, winds and ocean currents constantly transport heat poleward, where the extensive snow and ice cover influences its fate. As long as this highly reflective cover is intact and extensive, sunlight coming directly into the Arctic is mostly reflected back into space, keeping the Arctic cool and a good repository for the heat brought in from lower latitudes. But if the cover begins to melt and shrink, it will reflect less sunlight, and the Arctic will become a poorer repository, eventually warming the climate of the entire planet.

Figuring out just what will happen, however, is fraught with complications. The greatest of these stems from the intricate feedback systems that govern the climate in the Arctic. Some of these processes are positive, amplifying change and turning a nudge into a shove, and some are negative, behaving as a brake on the system and mitigating change.

Chief among these processes is the ice-albedo feedback, in which rising temperatures produce shorter winters and less extensive snow and ice cover, with ripple effects all the way back through the midlatitudes. Another feedback is associated with the large stores of carbon frozen into the Arctic in the form of peat. As the climate warms and this peat thaws, it could release carbon dioxide into the atmosphere and enhance warming over not just the Arctic but the whole globe—a phenomenon commonly referred to as greenhouse warming.

The key problem is that we don't fully understand how some of these feedback processes work in isolation, let alone how they interact. What we do know is that the Arctic is a complex system: change one thing, and everything else responds, sometimes in a counterintuitive way.

HEATING UP

The more we look, the more change we see. Arctic air temperatures have increased by 0.5 degree Celsius each decade over the past 30 years, with most of the warming coming in winter and spring. Proxy records (ice and peat cores, lake sediments), which tell us mostly about summer temperatures, put this recent warming in perspective. They indicate that late 20th- and early 21st-century temperatures are at their highest level in 400 years. The same records tell us that these high levels are the result of steady warming for 100 years as the Arctic emerged from the Little Ice Age, a frigid period that ended around 1850, topped off by a dramatic acceleration of the warming in the past half a century.

The recent temperature trends are mirrored in many other time series. One example is that Arctic and Northern Hemisphere river and lake ice has been forming later and melting earlier since the Little Ice Age. The total ice-cover season is 16 days shorter than it was in 1850. Near one of our homes (Sturm's) in Alaska, a jackpot of about $300,000 awaits the person who can guess the date the Tanana River will break up every spring.

The average winning date has gotten earlier by about six days since the betting pool was instituted in 1917. Higher-tech data—satellite images—show that the snow-free season in the Arctic has lengthened by several days each decade since the early 1970s. Similarly, the growing season has increased by as much as four days.

SHRINKING GLACIERS, THAWING PERMAFROST

There was nothing complex about my first research in Arctic climate change: march around a small glacier on Ellesmere Island, drill holes in the ice, insert long metal poles in the holes, measure them, come back a year later and see if more pole was showing.

We put in most of the pole network in the warm summer of 1982 and returned in 1983 to a very different world—week after week of cold, snow and fog. It looked like the start of a new ice age. Our plan had been to go back annually, but as so often happens, funding dried up, and my Arctic experiences became fond memories.

But memories sometimes get refreshed. In 2002 I got a call from an excited graduate student. He had revisited the glacier. It was rapidly wasting away. 1983 had been an anomaly. My stakes were there, except they were all lying on the surface of the ice. How deeply had I installed them? Did I still have my field notes? He need not have worried. There was my field book, dusty but safe in my bookcase. Now I'm going back to Ellesmere Island, to see what's left of the glacier that in 1983 seemed like such a permanent feature of the landscape but that I now realize may well die before I do.   —Mark C. Serreze

Arctic glaciers tell a striking tale as well. In Alaska, they have been shrinking for five decades, and more startlingly, the rate of shrinkage has increased threefold in the past 10 years. The melting glaciers translate into a rise in sea level of about two millimeters a decade, or 10 percent of the total annual rise of 20 millimeters. Determining the state of the much larger and more slowly changing Greenland Ice Sheet has been something of a Holy Grail for Arctic researchers. Older field and satellite results suggested that the ice sheet was exhibiting asymmetrical behavior—the west side thinning in a modest way and the east side remaining in balance. Recent satellite images indicate that the melt rate over the entire ice sheet has been increasing with time. The total area melting in a given summer has increased by 7 percent each decade since 1978, with last summer setting an all-time record. Winter snowfall appears insufficient to offset this heavy summer melt, so the sheet is shrinking.

The permafrost—the permanently frozen layer below the surface—is thawing, too. In a study published in 1986, researchers from the U.S. Geological Survey carefully logged temperature profiles in deep oil-exploration boreholes drilled through the permafrost of northern Alaska. When they extrapolated the profiles to the surface, they found an anomalous curvature that was best explained by a warming at ground level of two to four degrees C during the preceding few decades. More recently,

preliminary results suggest an additional increase of two to three degrees C has occurred since 1986. Because the Arctic winter lasts nine
months of the year, snow cover controls the thermal state of the ground
as much as air temperature does, so these borehole records almost certainly reflect a change in the amount and timing of winter precipitation
as well as an increase in temperature. More snow means thicker insula-
tion and therefore better protection for the ground from frigid winter
temperatures. Ground that is not chilled as much in the winter is primed
for more warming in the summer.

Regardless of why it is occurring, one thing is certain. Thawing permafrost is trouble. It can produce catastrophic failure of roads, houses and
other infrastructure. It is also implicated in another recently detected
change: over the past 60 years, the discharge of freshwater from Russian
rivers into the Arctic Basin has increased by 7 percent—an amount equivalent to roughly one quarter the volume of Lake Erie or three months of
the outflow of the Mississippi River. Scientists attribute the change partly
to greater winter precipitation and partly to a warming of the permafrost
and active layer, which they believe is now transporting more groundwater. This influx of freshwater could have important implications for global
climate: the paleo-record suggests that when the outflow of water from
the Arctic Basin hits a critical level of freshness, the global ocean circulation changes dramatically. When ocean circulation changes, climate does
as well, because the circulation system—essentially a set of moving rivers
of water in the ocean, such as the Gulf Stream—is one of the prime con-
veyors of heat northward toward the pole.

GREENING OF THE ARCTIC

The arctic land cover is also shifting. Based on warming experiments using greenhouses, biologists have known for some time that shrubs will
grow at the expense of the other tundra plants when the climate warms.
Under the same favorable growing conditions, the tree line will migrate
north. Researchers have been looking for these modifications in the real
world, but ecosystem responses can be slow. Only in the past few years, by
comparing modern photographs with ones taken 50 years ago, and by using satellites to detect the increasing amount of leaf area, have researchers been able to document that both types of transformations are under
way. As the vegetation alters, so does the role of the Arctic in the global
carbon cycle. Vast stores of carbon in the form of peat underlie much of
the tundra in Alaska and Russia, evidence that for long periods Arctic
tundra has been a net carbon sink; about 600 cubic miles of peat are cur-

rently in cold storage. In recent years, warming has produced a shift: the Arctic now appears to be a net source of carbon dioxide. The change is subtle but troubling because carbon dioxide and methane constitute the primary greenhouse gases in the atmosphere, returning heat to the earth instead of allowing it to escape into space.

Warmer winters have driven some of the shift. When the air is warmer, more precipitation falls from the sky, some of it coming as snow. The thicker snow holds more warmth in the earth, resulting in a longer period during which the tundra is releasing carbon dioxide. But as the tundra becomes shrubbier and as the soil becomes drier in the summer as a result of higher temperatures, the balance could swing back the other way, because plants, particularly woody ones, will fix more carbon and lock it back into the Arctic ecosystem. The most recent studies suggest, in fact, that the magnitude and direction of the Arctic carbon balance depend on the time span that we are examining, with the response varying as the plants adapt to the new conditions.

MELTING SEA ICE

"This sea ice is ridiculously thin," I thought as I broke through the ice for the second time that morning in August 1998. There was no real danger, now that personal flotation devices had become the the de rigueur fashion accessory, but the thin ice was troubling for other reasons.

My journey to this place, 600 miles from the North Pole, had begun 10 months earlier on board the icebreaker *Des Groseilliers,* which we had intentionally frozen into the pack to begin a yearlong drift. Our mission was to study ice-albedo and cloud-radiation feedbacks. When we started the journey, I was surprised at how thin the ice was. Now, after a much longer than expected summer melt season, it was thinner still, even though we had been drifting steadily north. I was uncertain which would come first: the end of the summer or the end of the ice. Little did I know that this summer the record for minimum ice cover was being set throughout the entire western Arctic Ocean. Unfortunately for the long-term survival of the ice pack, it was a record that was easily broken in 2002.   —Donald K. Perovich

Of all the changes we have catalogued, the most alarming by far has been the reduction in the Arctic sea-ice cover. Researchers tracking this alteration have discovered that the area covered by the ice has been decreasing by about 3 percent each decade since the advent of satellite records in 1972. This rate might be low for a financial investment, but where time is measured in centuries or millennia, it is high. With the sea ice covering an area approximately the size of the U.S., the reduction per decade is equivalent to an area the size of Colorado and New Hampshire combined, the home states of two of us (Perovich and Serreze). The change in the thickness of the ice (determined from submarines) is even more striking: as much as 40 percent lost in the past few decades. Some climate models suggest that by 2080 the Arctic Ocean will be ice-free in summer.

The melting sea ice does not raise sea level as melting glaciers do, because the ice is already floating, but it is alarming for two other reasons. Locally, the demise of the sea ice leads to the loss of a unique marine ecosystem replete with polar bears, seals and whales. Globally, an ice-free Arctic Ocean would be the extreme end point of the ice-albedo feedback—far more solar radiation would be absorbed, warming not just the Arctic but eventually every part of the earth.

The shrinking sea-ice cover has not escaped the attention of businesspeople, tourists and politicians. Serious discussions have been under way about the feasibility of transporting cargo via Arctic waters—including through the fabled Northwest Passage, now perhaps close to being a practical shipping route because of climate change. Roald Amundsen, the redoubtable Norwegian polar explorer, took more than three years to complete the first transit of the passage in 1906, when the Arctic was still under the influence of the Little Ice Age. Many explorers before him had died trying to make the journey. In the past few years, however, dozens of ships have completed the route, including Russian icebreakers refurbished for the tourist trade. These events would have been unimaginable, even with icebreakers, in the more intense ice conditions of 100 years ago.

IS GREENHOUSE WARMING THE CULPRIT?

This inventory of startling transformation in the Arctic inevitably raises the question of whether we are still emerging from the Little Ice Age or whether something quite different is now taking place. Specifically, should we interpret these changes as being caused by the increased concentration of atmospheric greenhouse gases overriding a natural temperature cycle? Or are they part of a longer-than-expected natural cycle?

The intricate web of feedback interactions renders this question exceedingly complicated—and we don't know enough yet to answer it unequivocally. But we know enough to be very worried.

Whatever is causing the melting and thawing now wracking the Arctic, these modifications have initiated a cascade of planetwide responses that will continue even if the climate were suddenly and unexpectedly to stop warming. Imagine the climate as a big, round rock perched on uneven terrain. The inventory tells us that the rock has been pushed a little—either by a natural climate cycle or by human activity—and has started to roll. Even if the pushing stops, the rock is going to keep rolling. When it finally does stop, it will be in a completely different place than before.

To cope with the constellation of changes in the Arctic in a concerted fashion and to develop an ability to predict what will happen next rather than just react to it, several federal agencies have begun to coordinate their Arctic research in a program called SEARCH (Study of Environmental Arctic Change). Early results give some promise for success in teasing out the linkages among the tightly coupled systems that shape the climate of the Arctic and thus the earth. A recent discovery about the patterns of wind circulation, for example, helps to explain previously puzzling spatial patterns of increasing temperature. Equally important, high-quality records of climate change now extend back 30 to 50 years.

Soon these records and other findings should allow us to determine whether the Arctic transformation is a natural trend linked to emergence from the Little Ice Age or something more ominous. Our most difficult challenge in getting to that point is to come to grips with how the various feedbacks in the Arctic system interact—and to do so quickly.

FURTHER READING

YEAR ON ICE GIVES CLIMATE INSIGHTS. D. K. Perovich et al. in *EOS, Transactions of the American Geophysical Union*, Vol. 80, No. 481, pages 485–486; 1999.

GLOBAL WARMING AND TERRESTRIAL ECOSYSTEMS: A CONCEPTUAL FRAMEWORK FOR ANALYSIS. G. R. Shaver et al. in *BioScience*, Vol. 50, No. 10; 2000.

OBSERVATIONAL EVIDENCE OF RECENT CHANGE IN THE NORTHERN HIGH-LATITUDE ENVIRONMENT. Mark Serreze et al. in *Climatic Change*, Vol. 46, pages 159–207; 2000.

THE SURFACE HEAT BUDGET OF THE ARCTIC OCEAN (SHEBA). Special section in *Journal of Geophysical Research*, Vol. 107, No. 15; October 2002.

SEARCH Web site: http://psc.apl.washington.edu/search/

NOAA Arctic Climate Change Web site: www.ngdc.noaa.gov/paleo/sciencepub/front.htm

ILLUSTRATION CREDITS

Pages 18, 20, 21: Ian Worpole. Pages 38 and 40: Johnny Johnson. Page 47: Heidi Noland. Pages 50–51: David Fierstein. Page 54: Heidi Noland. Page 78: courtesy of the SeaWiFS, NASAS Goddard Space Flight Center and Orbimage. Page 81: David Fierstein. Page 97: from Fernandez de Oviedo's *History of the Indies* (1535). Pages 99 and 102: Roberto Osti. Page 104: Ian Worpole. Pages 108 and 110–11: William C. Ober. Page 113: Jana Brenning and William C. Ober. Page 118: Laurie Grace. Page 130 Cleo Vilett. Page 131: Nina Finkel. Page 133: Nina Finkel (graph); Cleo Vilett (illustration). Pages 136 and 141: Laurie Grace and Roberto Osti. Page 142: Laurie Grace. Page 149: Roberto Osti. Page 151: Jennifer C. Christiansen; sources: U.N. Food and Agriculture and *Shrimp News International*. Pages 156 and 158: William F. Haxby. Page 160: Laurie Grace. Page 166: map by Laurie Grace; sources: Marine Aquarium Council and International Marinelife Alliance. Pages 174, 178, 188, and 190: Laurie Grace. Page 193: Johnny Johnson. Page 198: Laurie Grace. Page 199: Jennifer C. Christiansen. Page 205: Aries Galindo and Laurie Grace. Page 210: David Fierstein. Pages 217 and 218: Jennifer C. Christiansen. Pages: 237, 240, and 241: Laurie Grace. Page 252: Johnny Johnson. Pages 256 and 258: John de Santis. Page 268: Slim Films. Page 270: Laurie Grace. Page 276: Alison Kendall (illustration); David Hawkins, Natural Resources Defense Council (concept); Gregg Marland, Oak Ridge National Laboratory (past data). Pages 276 and 284: Barry Ross. Pages 289 and 290: Jana Brenning. Page 296: Jennifer C. Christiansen; source: J. R. Petit et al. in *Nature*, vol. 399, pages 429–436, June 3, 1999. Pages 298, 302, and 305: Jennifer C. Christiansen; source: James Hansen.